Lecture Notes in Statistics

Vol. 1: R. A. Fisher: An Appreciation. Edited by S. E. Fienberg and D. V. Hinkley. XI, 208 pages, 1980.

Vol. 2: Mathematical Statistics and Probability Theory. Proceedings 1978. Edited by W. Klonecki, A. Kozek, and J. Rosiński. XXIV, 373 pages, 1980.

Vol. 3: B. D. Spencer, Benefit-Cost Analysis of Data Used to Allocate Funds. VIII, 296 pages, 1980.

Vol. 4: E. A. van Doorn, Stochastic Monotonicity and Queueing Applications of Birth-Death Processes. VI, 118 pages, 1981.

Vol. 5: T. Rolski, Stationary Random Processes Associated with Point Processes. VI, 139 pages, 1981.

Vol. 6: S. S. Gupta and D.-Y. Huang, Multiple Statistical Decision Theory: Recent Developments. VIII, 104 pages, 1981.

Vol. 7: M. Akahira and K. Takeuchi, Asymptotic Efficiency of Statistical Estimators. VIII, 242 pages, 1981.

Vol. 8: The First Pannonian Symposium on Mathematical Statistics. Edited by P. Révész, L. Schmetterer, and V. M. Zolotarev. VI, 308 pages, 1981.

Vol. 9: B. Jørgensen, Statistical Properties of the Generalized Inverse Gaussian Distribution. VI, 188 pages, 1981.

Vol. 10: A. A. McIntosh, Fitting Linear Models: An Application on Conjugate Gradient Algorithms. VI, 200 pages, 1982.

Vol. 11: D. F. Nicholls and B. G. Quinn, Random Coefficient Autoregressive Models: An Introduction. V, 154 pages, 1982.

Vol. 12: M. Jacobsen, Statistical Analysis of Counting Processes. VII, 226 pages, 1982.

Vol. 13: J. Pfanzagl (with the assistance of W. Wefelmeyer), Contributions to a General Asymptotic Statistical Theory. VII, 315 pages, 1982.

Vol. 14: GLIM 82: Proceedings of the International Conference on Generalised Linear Models. Edited by R. Gilchrist. V, 188 pages, 1982.

Vol. 15: K. R. W. Brewer and M. Hanif, Sampling with Unequal Probabilities. IX, 164 pages, 1983.

Vol. 16: Specifying Statistical Models: From Parametric to Non-Parametric, Using Bayesian or Non-Bayesian Approaches. Edited by J. P. Florens, M. Mouchart, J. P. Raoult, L. Simar, and A. F. M. Smith. XI, 204 pages, 1983.

Vol. 17: I. V. Basawa and D. J. Scott, Asymptotic Optimal Inference for Non-Ergodic Models. IX, 170 pages, 1983.

Vol. 18: W. Britton, Conjugate Duality and the Exponential Fourier Spectrum. V, 226 pages, 1983.

Vol. 19: L. Fernholz, von Mises Calculus For Statistical Functionals. VIII, 124 pages, 1983.

Vol. 20: Mathematical Learning Models – Theory and Algorithms: Proceedings of a Conference. Edited by U. Herkenrath, D. Kalin, W. Vogel. XIV, 226 pages, 1983.

Vol. 21: H. Tong, Threshold Models in Non-linear Time Series Analysis. X, 323 pages, 1983.

Vol. 22: S. Johansen, Functional Relations, Random Coefficients and Nonlinear Regression with Application to Kinetic Data. VIII, 126 pages. 1984.

Vol. 23: D. G. Saphire, Estimation of Victimization Prevalence Using Data from the National Crime Survey. V, 165 pages. 1984.

Vol. 24: T. S. Rao, M. M. Gabr, An Introduction to Bispectral Analysis and Bilinear Time Series *Models. VIII, 280 pages, 1984.*

Vol. 25: Time Series Analysis of Irregularly Observed Data. Proceedings, 1983. Edited by E. Parzen. VII, 363 pages, 1984.

ctd. on inside back cover

Lecture Notes in Statistics

Edited by D. Brillinger, S. Fienberg, J. Gani, J. Hartigan, and K. Krickeberg

37

Advances in Order Restricted Statistical Inference

Proceedings of the Symposium on Order Restricted Statistical Inference held in Iowa City, Iowa, September 11–13, 1985

Edited by R. Dykstra, T. Robertson and F. T. Wright

Editors

Richard Dykstra
Tim Robertson
University of Iowa, Department of Statistics and Actuarial Sciences
Iowa City, Iowa 52242, USA

Farroll T. Wright
University of Missouri, 323 Mathematics-Computer Science
Rolla, Missouri 65401, USA

AMS Subject Classification: 62-06, 62C10, 62F03, 62F04, 62F10, 62F35, 62G05, 62G10

ISBN-13: 978-0-387-96419-5 e-ISBN-13: 978-1-4613-9940-7
DOI: 10.1007/978-1-4613-9940-7

Softcover reprint of the hardcover 1st edition 1986

2147/3140-543210

PREFACE

With support from the University of Iowa and the Office of Naval Research, a small conference on order restricted inference was held at the University of Iowa in Iowa City in April of 1981. There were twenty-one participants, mostly from the midwest, and eleven talks were presented. A highlight of the conference was a talk by D. J. Bartholomew on, "Reflections on the past and thoughts about the future."

The conference was especially valuable because it brought together researchers who were thinking about related problems. A small conference on a limited topic is one of the best ways to stimulate research and facilitate collaboration. Because of the success of the first conference, a second conference was organized and held in September of 1985. This second conference was made possible again by support from the Office of Naval Research under Department of the Navy Contract N00014-85-0161 and the University of Iowa. There were thirty-five participants and twenty presentations on a wide variety of topics dealing with order restricted inference at the second conference. This volume is a collection of fourteen of those presentations.

By collecting together and organizing the fundamental results in order restricted inference in *Statistical Inference under Order Restrictions*, R. E. Barlow, D. J. Bartholomew, J. M. Bremner and H. D. Brunk have done much to stimulate research in this area, and so we wish to express our gratitude to them first. We also wish to express our appreciation to those who attended the Second Conference on Order Restricted Statistical Inference for their enthusiastic participation, their cooperation in the preparation of this volume and, of course, for their contributions to this area of research. The First Conference on Order Restricted Inference has already had a definite effect on the development of this area and we fully expect the impact of this second conference to be even greater.

This conference would not have been possible without the funding provided by the Office of Naval Research. We gratefully acknowledge this support and wish to thank L. D. Broemeling and E. J. Wegman for their assistance in arranging it. The conference was jointly sponsored by the Department of Statistics and Actuarial Science at the University of Iowa. We wish to express our appreciation to John J. Birch, Acting Chairman, for his help in arranging for the conference and Mrs. Ada Burns for her able help in preparing the manuscripts for publication. The staff at Springer-Verlag has also been most helpful in arranging for the publication of this volume. Finally, allow us to say thanks to those who helped by refereeing the manuscripts submitted. This volume owes a great deal to the assistance of these anonymous editorial collaborators.

CONTENTS

PREFACE III

CONFERENCE PARTICIPANTS VII

INTRODUCTION. 1

Bayesian and Maximum Likelihood Approaches to Order Restricted Inference for Models from Ordinal Categorical Data

Alan Agresti and Christy Chuang. 6

A Method of Finding Projections Onto the Intersection of Convex Sets in Hilbert Spaces

James P. Boyle and Richard Dykstra 28

Isotonic M-Estimation

Sue Luergans 48

An Application of Isotonic Regression to Multivariate Density Estimation

Thomas W. Sager. 69

Least-Squares Regression Under Convexity and Higher Order Difference Constraints with Applications to Software Reliability

Douglas Miller and Ariela Sofer. 91

Dose-Response Analysis under Unimodality of Response-to-Dose

Richard L. Schmoyer. 125

On Nonparametric Tests for Ordered Alternatives in Two-Way Layouts

Z. Govindarajulu and S. H. Mansouri-Ghiassi. 153

A Class of Distribution-Free Tests for Testing Homogeneity of Variances Against Ordered Alternatives

Subhash C. Kochar and R. P. Gupta. 169

Robustness of $\bar{\chi}^2$ and $\bar{E}^2$: A Case for a Ranked Test in Ordered Restricted Inference

Michael Schell and Bahadur Singh. 184

Multiple Contrast Tests for Testing Against a Simple Tree Ordering

Hari Mukerjee, Tim Robertson, and F. T. Wright 203

On Testing Symmetry and Unimodality
Tim Robertson. 231

On Tests Under Order Restrictions in Generalized Homogeneity and Reduction of Dimensionality
Ashis SenGupta 249

Power Series Approximations to the Null Distributions of Some Chi-Bar-Square Statistics
Bahadur Singh and F. T. Wright 257

Conditional Tests With an Order Restriction as a Null Hypothesis
Peter C. Wollan and Richard Dykstra. 279

Conference Participants

September 11-13, 1985

Alan AGRESTI, Department of Statistics, University of Florida

Roger BERGER, Department of Statistics, North Carolina State University

James P. BOYLE, Department of Statistics & Actuarial Science, University of Iowa

Richard DYKSTRA, Department of Statistics & Actuarial Science, University of Iowa

Luis ESCOBAR, Experimental Statistics, Louisiana State University

Carol FELTZ, Applied Mathematics & Statistics Department, AT&T

A. M. FINK, Department of Mathematics, Iowa State University

Z. GOVINDARAJULU, Department of Statistics, University of Kentucky

John HEWETT, Department of Statistics, University of Missouri at Columbia

Robert V. HOGG, Department of Statistics & Actuarial Science, University of Iowa

Robert KELLY, Department of Statistics & Actuarial Science, University of Iowa

S. C. KOCHAR, Department of Mathematics & Statistics, Dalhousie University, Canada

Purushottam LAUD, Department of Mathematics & Statistics, Northern Illionis University

Charles LEE, Department of Mathematics & Statistics, Memorial University of Newfoundland, Canada

Tze-San LEE, Department of Mathematics, Western Illinois University

Jon LEMKE, Department of Preventive Medicine, University of Iowa

Sue LEURGANS, Department of Statistics, Ohio State University

Douglas MILLER, Department of Operations Research, George Washington University

Hari MUKERJEE, Division of Statistics, University of California at Davis

Jong-Shi PANG, School of Management & Administration, University of Texas at Dallas

Carolyn PILLERS, Department of Statistics and Actuarial Science, University of Iowa

Richard F. RAUBERTAS, Biometry & Field Studies Branch, IRP, National Institute of Health

Tim ROBERTSON, Department of Statistics & Actuarial Science, University of Iowa.

Ralph RUSSO, Department of Statistics & Actuarial Science, University of Iowa

Tom SAGER, Department of General Business, University of Texas at Austin

Michael SCHELL, Biostatistics Division, St. Jude Children's Hospital

Richard L. SCHMOYER, Mathematics & Statistics Research Department, Oak Ridge National Laboratories

Ashis SENGUPTA, Indian Statistical Institute, INDIA

Bahadur SINGH, Department of Mathematical Sciences, Memphis State University

Ariela SOFER, System Engineering Department, George Mason University

Shashikala SUKHATME, Department of Statistics, Iowa State University

Giles WARRACK, Department of Mathematics, University of North Carolina at Greensboro

Peter WOLLAN, Department of Statistics, Michigan Technological University

F. T. WRIGHT, Department of Mathematics, University of Missouri at Rolla

INTRODUCTION

Prior information regarding a statistical model frequently constrains the shape of the parameter set and can often be quantified by placing inequality constraints on the parameters. For example, the expected response or the probability of a specific response may increase or decrease with the treatment level; a regression function may be nondecreasing or convex or both; the failure rate of a component may increase as it ages; or the treatment response may stochastically dominate the control. The fact that utilization of such ordering information increases the efficiency of procedures developed for statistical inference is well documented. The one-tailed two-sample t-test provides a familiar example in which the procedure which utilizes the prior information (the one-sided test) dominates procedures which ignore this information.

As is often the case in science, several researchers began work, independently, on related problems in order restricted inference during the early fifties. The period from 1950 through 1970 was an active period of research in order restricted inference and the monograph by Barlow, Bartholomew, Bremner and Brunk which was published in 1972 lists over 250 works related to this topic. Of course many of these references only touch on order restricted inference. However, this list does give one a feeling of the breadth and depth of the subject. Many prominent statisticians have contributed to this area. Moreover, order restricted inference touches on some of the most active areas of research today such as robustness, reliability, density estimation, computational procedures, asymptotic theory, and the mathematics of shape and form. One can obtain a sense of the history of this work by reading the Complements Sections at the end of each chapter of Barlow et al. (1972). In our opinion, a few of the landmark works of this period are the following: Chernoff (1954) found the asymptotic distribution of the likelihood ratio statistic for testing whether a parametric vector is located on one side of a hyperplane. This seems to be one of the first developments of a chi-bar-square distribution. Ayer, Brunk, Ewing, Reid, and Silverman (1955) studied maximum likelihood estimates of completely ordered parameters in a bioassay setting. Brunk (1955) and van Eeden (1956) considered the estimation of parameters subject to the restrictions imposed by a partial order on their index set. For the one-way analysis of variance, Bartholomew (1959) developed the likelihood ratio test of homogeneity with the alternative constrained by order restrictions. Chacko (1963) considered the analogous problem in a multinomial setting and the Jonckheere-Terpstra test (cf. Terpstra (1952) and Jonckheere (1954)) is a distribution free competitor to Bartholomew's chi-bar-square test. The estimation of a distribution with increasing failure rate was studied by Grenander (1956), Marshall and Proschan (1965), and Barlow and Van Zwet (1970). Estimation of restricted densities was studied by Grenander (1956), Robertson (1967), Prakasa Rao (1969) and Wegman (1970). Kudô (1963) developed much of the early theory for tests involving order restrictions in a multivariate normal setting and Perlman (1969) generalized existing multivariate results and studied properties of power functions of some of the chi-bar-square tests. The monograph by Barlow et al. (1972) contains an excellent exposition of the key concepts in order restricted inference up to the early seventies.

More recent research in order restricted inference has extended the work begun in those early years and has provided new algorithms for making the early work more accessible in the practice of statistics. Theory and methods for estimates and tests involving more complex order restrictions have been developed and the theory of duality (cf. Barlow and Brunk (1972)) has provided insights into new problems. The area of categorical data analysis is a fertile field for order restricted research and several problems have recently been solved. Testing situations in which both the null and alternative hypotheses involve order constraints are being studied. Study of the level probabilities involved in the chi-bar-square distribution continues but there remain order restrictions of practical importance for which little is known about these level probabilities. Because of the complexity of the likelihood ratio procedures, alternative approaches, such as multiple contrast tests and conditional tests, are being studied. It has been known for some time that estimates and tests based upon the likelihood function can be inadmissible, but recent work has produced alternative estimates and tests which have improved error and power characteristics.

Along with the continuation of the work mentioned above, there are exciting challenges in the development of Bayesian procedures and interval estimates which utilize ordering information. There is also the need to make use of the theory developed in order restricted inference in other areas of statistical methodology. Order restricted inference contains many powerful statistical techniques and there is a need for these procedures to receive more widespread use in the practice of statistics.

The papers collected together in this volume are representative of recent advances in order restricted inference. Contributions to Bayesian methods, multivariate analysis, nonparametric and robust theory, algorithms and approximations for estimates and tests are presented, and new testing situations and applications are explored. The editors are proud to be associated with the fourteen papers in this volume and we thank the authors for their contributions.

The Editors

REFERENCES

Ayer, M., Brunk, H.D., Ewing, G.M., Reid, W.T. & Silverman, E. (1955). An empirical distribution function for sampling with incomplete information. *Ann. Math. Statist.* **26**, 641-647.

Barlow, R.E., Bartholomew, D.J., Bremner, J.M. & Brunk, H.D. (1972). *Statistical Inference under Order Restrictions*. Wiley: New York.

Barlow, R.E. & Brunk, H.D. (1972). The isotonic regression problem and its dual. *J. Amer. Statist. Assoc.* **67**, 140-147,

Barlow, R.E. & van Zwet, W.R. (1970). Asymptotic properties of isotonic estimators for the generalized failure rate function. Part I: Strong consistency. In M.L. Puri (Ed.), *Nonparametric Techniques in Statistical Inference*. Cambridge University Press, 159-173.

Bartholomew, D.J. (1959). A test of homogeneity for ordered alternatives. *Biometrika* **46**, 36-48.

Brunk, H.D. (1955). Maximum likelihood estimates of monotone parameters. *Ann. Math. Statist.* **26**, 607-616.

Chacko, V.J. (1963). Testing homogeneity against ordered alternatives. *Ann. Math. Statist.* **34**, 945-956.

Chernoff, H. (1954). On the distribution of the likelihood ratio. *Ann. Math. Statist.* **25**, 573-578.

Eeden, C. van (1956). Maximum likelihood estimation of ordered probabilities. *Proc. K. ned. Akad. Wet.*(A), 59/*Indag. math.* **18**, 444-455.

Grenander, U. (1956). On the theory of mortality measurement. Part II. *Skand. Akt.* **39**, 125-153.

Jonckheere, A.R. (1954). A distribution-free *k*-sample test against ordered alternatives. *Biometrika* **41**, 133-145.

Kudô, A. (1963). A multivariate analogue of the one-sided test. *Biometrika* 50, 403-418.

Marshall, A.W. & Proschan, F. (1965). Maximum likelihood estimation for distributions with monotone failure rate. *Ann. Math. Statist.* 36, 69-77.

Perlman, M.D. (1969). One-sided problems in multivariate analysis. *Ann. Math. Statist.* **40**, 549-567 (for corrections to the above paper, see *Ann. Math. Statist.* **42**, 1777).

Prakasa Rao, B.L.S. (1969). Estimation of a unimodal density. *Sankhya (A)* **3**, 23-36.

Robertson, T. (1967). On estimating a density which is measurable with respect to a σ-lattice. *Ann. Math. Statist.* **38**, 482-493.

Terpstra, T.J. (1952). The asymptotic normality and consistency of Kendall's test against trend when ties are present in one ranking. *Proc. Sect. Sci. K. ned. Akad. Wet. (A)* 55/*Indag. math.* **14**, 327-333.

Wegman, E.J. (1970). Maximum likelihood estimation of a unimodal density function. *Ann. Math. Statist.* **41**, 457-471.

BAYESIAN AND MAXIMUM LIKELIHOOD APPROACHES TO ORDER-RESTRICTED INFERENCE FOR MODELS FOR ORDINAL CATEGORICAL DATA[1]

Alan Agresti
University of Florida

Christy Chuang
The Upjohn Company

AMS 1980 subject classifications: 62H17, 62A15.

Key words and phrases: *Isotonic regression, likelihood-ratio dependence, loglinear models, odds ratio, order statistics, row effects model, row and column effects model, uniform prior.*

ABSTRACT

A class of association models for contingency tables has parameters that are sometimes interpreted as category scores. For classifications having ordered categories, it is often reasonable to assume that the score parameters have a corresponding ordering. This article proposes order-restricted estimates of score parameters in these models. For these estimates, the local log odds ratios have uniform sign. For the Bayesian approach proposed here, prior distributions can induce the order restriction, and prior beliefs reflecting strong association have the effect of moving the estimates away from the boundary of the restricted parameter space. The order-restricted maximum likelihood solution is obtained in the limit as the prior standard deviation for the strength of association parameter grows unboundedly.

1. INTRODUCTION. Suppose that a sample of n subjects is cross classified by two ordinal variables X and Y into a $r \times c$ table.

---

1. Research partially supported by grant R01 GM33210 of the National Institutes of Health for Dr. Agresti and grant CA11198 of the National Cancer Institute for Dr. Chuang.

Denote the expected frequencies in that table by $\{m_{ij}\}$. The local odds ratios

$$\theta_{ij} = m_{ij}m_{i+1,j+1}/m_{i,j+1}m_{i+1,j}, \quad 1 \le i \le r-1, \quad 1 \le j \le c-1,$$

are useful for describing properties of models for the association between X and Y. Perhaps the simplest and most useful association models are special cases of the model

$$(1.1) \qquad \log m_{ij} = \mu + \lambda_i^X + \lambda_j^Y + \beta\mu_i\nu_j$$

proposed by Goodman (1979, 1981a). For the special case $\{\mu_i = i\}$, $\{\nu_j = j\}$, this is the *uniform association* (U) model, for which $\{\log \theta_{ij} = \beta\}$. In the general case in which the $\{\mu_i\}$ and $\{\nu_j\}$ are unspecified parameters,

$$\log \theta_{ij} = \beta(\mu_{i+1}-\mu_i)(\nu_{j+1}-\nu_j),$$

and the model is referred to as the *multiplicative row and column effects* (or RC) model. It is referred to as the *row effects* (R) model when the $\{\mu_i\}$ are parameters and the $\{\nu_j\}$ are fixed, monotone scores, and the *column effects* (C) model when the $\{\nu_j\}$ are parameters and the $\{\mu_i\}$ are fixed, monotone scores. One reason for the importance of this structural form is that it can be regarded as a discrete version of a family of distributions that includes the bivariate normal distribution. Model (1.1) tends to fit very well when there is an underlying continuous distribution that is approximately of that form (see Goodman, 1981b).

In model (1.1) we will take $\sum \lambda_i^X = \sum \lambda_j^Y = 0$. Location or scale changes in the "score" parameters $\{\mu_i\}$ or $\{\nu_j\}$ do not alter the basic form of the R, C or RC models. Here we adopt the scaling $\mu_1 = 1$, $\mu_r = r$, $\nu_1 = 1$, $\nu_c = c$ to make the scores comparable in value to the fixed integer scores in the U model. Since $\sum\sum \log \theta_{ij} = (\mu_r - \mu_1)(\nu_c - \nu_1)\beta$, this scaling gives the simple interpretation for β as the average of the $(r-1)(c-1)$ local log odds ratios. This scaling is inadequate if a model holds with $\mu_1 = \mu_r$ or $\nu_1 = \nu_c$, but the analyses we propose are not appropriate for such cases.

The standard fits of these models do not fully utilize the ordinal nature of the variables, in the sense that the same $\{\hat{m}_{ij}\}$ and goodness-of-fit statistics are obtained if levels of variables having parameter scores are permuted in any way. Hence, the regular solutions do not necessarily lead to inferences that reflect the types of departures from independence expected with ordinal data. In many applications, for instance, we expect the orderings of the categories to be manifested in an association that is monotonic, in some sense. One possible operationalization of monotonicity is the condition that the $\{\log \theta_{ij}\}$ are uniformly nonnegative (or uniformly nonpositive). In a more general context, Lehmann (1966) referred to this condition as *positive* (or *negative*) *likelihood-ratio dependence*. For model (1.1), likelihood-ratio dependence is equivalent to the constraints for the score parameters,

$$(1.2) \qquad \mu_1 \le \mu_2 \le \cdots \le \mu_r \quad \text{and} \quad \nu_1 \le \nu_2 \le \cdots \le \nu_c .$$

In this article we give a Bayesian approach to estimating the parameters in model (1.1). The prior distribution in this approach implies the belief that the local log odds ratios are uniformly of one sign. This results in a posterior fit of the model for which the score estimates have the same ordering as do the categories of the ordinal variables (i.e., they satisfy (1.2)). An order-restricted maximum likelihood (ML) solution is obtained in the limit as the prior standard deviation for an association parameter grows unboundedly.

2. ORDER-RESTRICTED MAXIMUM LIKELIHOOD SOLUTIONS. Let $\underset{\sim}{x} = (x_{11}, \cdots, x_{rc}) = n\underset{\sim}{p}$ denote the observed cell counts, with $n = \sum\sum x_{ij}$. We assume that $\underset{\sim}{x}$ has a multinomial distribution with probabilities $\{\pi_{ij}\}$ corresponding to expected cell frequencies $\{m_{ij} = n\pi_{ij}\}$ satisfying (1.1). The order-restricted maximum likelihood (ML) solution maximizes $\Pi\Pi\, m_{ij}^{x_{ij}}$ subject to $1 = \mu_1 \leq \cdots \leq \mu_r = r$ for the R model and in addition $1 = \nu_1 \leq \cdots \leq \nu_c = c$ for the RC model. These solutions are discussed by Agresti, Chuang and Kezouh (1986), and proofs of the results quoted in this section can be found there. The most complete results hold for the R (or C) model. Suppose that the order-restricted estimates $\{\hat{\mu}_i\}$ for that model satisfy

$$\hat{\mu}_1 = \cdots = \hat{\mu}_{r_1} < \hat{\mu}_{r_1+1} = \cdots = \hat{\mu}_{r_2} < \cdots < \hat{\mu}_{r_{a-1}+1} = \cdots = \hat{\mu}_{r_a}$$

and let $R_k = \{r_{k-1}+1, \cdots, r_k\}$, $k = 1, \cdots, a$, with $r_0 = 0$. Using an argument analogous to that in the Appendix for the Bayes solution, it can be shown that the likelihood equations are

$$\hat{m}_{i+} = x_{i+}, \quad i = 1, \cdots, r, \tag{2.1a}$$

$$\hat{m}_{+j} = x_{+j}, \quad j = 1, \cdots, c, \tag{2.1b}$$

$$\sum_{R_k} \sum_j \upsilon_j \hat{m}_{ij} = \sum_{R_k} \sum_j \upsilon_j x_{ij}, \quad k = 1,\cdots,a, \tag{2.2}$$

where "+" symbolizes summation over the corresponding index.

The likelihood equations for the ordinary ML solutions are (2.1)-(2.2) with $R_k = \{k\}$, $k = 1,\cdots,r$. Thus, these equations give as a special case the likelihood equations for the R model fitted to the collapsed table in which the rows in each of $R_1,\cdots,R_a$ are combined. The order-restricted solution is therefore the same as the ordinary ML solution for the appropriately collapsed table. If the ordinary ML estimates are monotone, then the order-restricted ML estimates are identical to them. Analogous remarks apply to the RC model.

The R and C models are loglinear, so the log likelihoods under these models are concave and solutions are well-defined. The RC model is not loglinear, and we know of no algorithm that is guaranteed to give convergence to the ordinary ML solution, much less an order-restricted one. For the R (or C) model fitted with an additional constraint about the sign of the association (say, $\beta > 0$), the order-restricted solution is unique, and necessary and sufficient conditions can be given for determining the partition $\{R_k\}$ for which equations (2.1)-(2.2) give that solution:

PROPERTY A. *A partition* $\{R_k\}$ *gives the order-restricted ML solution if and only if it gives ordered estimates and, for every refinement* $\{R_1,\cdots,R_{i1},R_{i2},\cdots,R_a\}$, $i = 1,\cdots,a$, *where* $R_i = R_{i1} \cup R_{i2}$, *the ordinary ML solution violates the order constraint between sets* R_{i1} *and* R_{i2}.

PROPERTY B. *The partition* $\{R_k\}$ *for the order-restricted* ML *solution is identical to the partition of level sets obtained in using the pooling adjacent violators algorithm to obtain the regression of the sample row means* $\{\sum_j \nu_j x_{ij}/x_{i+},\ i = 1,\cdots,r\}$ *in the class of functions isotonic with respect to the simple order on the rows. In this isotonic regression, the row marginal totals are used as the weights.*

For the constraint $\beta < 0$, the correspondence is with the isotonic regression of row means with respect to the reverse of the simple order.

The order-restricted ML solution can be computed using a general-purpose program for maximizing a function subject to linear inequality constraints. Optimization methods used for obtaining the maximum are discussed in detail in Gill et al. (1981, Chap. 5). We have used the EO4UAF subroutine from the NAG library (1984) to obtain order-restricted solutions for these models. A sequential augmented Lagrangian method is used, the maximization subproblems involved being solved by a quasi-Newton method. Speed of convergence is considerably enhanced by supplying good initial estimates. For these, one could use order-restricted adjustments to the ordinary estimates for the model, which can be obtained using the EO4KAF subroutine in the NAG library. We have also used the BMDP-3R (Dixon 1979) nonlinear regression program to obtain order-restricted solutions, by using one of its options to form G^2 as the recognized loss function. Standard software that can handle the row effects model (see Agresti 1984, Appendix D) can also be used to obtain the order-restricted ML solution, through an iterative process of fitting the model to collapsed tables and checking the condition in Property A for lesser collapsings, or by first performing isotonic regression on the row means to determine directly the proper partition.

The goodness of fit of the order-restricted row effects model can be decomposed into the goodness of fit of independence models to the $\{R_k\}$ plus the goodness of fit of the row effects model to the collapsed table in which each set of rows R_k is combined into a single row. Specifically, let $G^2(R^*) = 2 \sum \sum x_{ij} \log(x_{ij}/\hat{m}_{ij})$ denote the likelihood-ratio statistic for the order-restricted fit for the original table, let $G^2(R')$ denote the fit of the R model for the collapsed table, let $G^2(I)$ and $G^2(I')$ denote the fit of the independence model to the original and collapsed tables, respectively, and let $G^2(I_k)$ denote the fit of the independence model to the set of rows R_k, $k = 1,\cdots,a$. Then, the following result holds:

PROPERTY C.
$$G^2(R^*) = G^2(R') + \sum G^2(I_k)$$
$$= G^2(R') + G^2(I) - G^2(I').$$

Also, it is shown in Agresti et al. (1986) that the difference between G^2 values for the order-restricted ML fit and for the regular ML fit is algebraically identical to a statistic for (a priori) testing of equality of certain score parameters. Analogous results are also given there for the RC model.

3. ORDER-RESTRICTED BAYES SOLUTION FOR RC MODEL. For the Bayes approach, we first consider the RC model, for which both the $\{\mu_i\}$ and $\{\nu_j\}$ are parameters. A priori, we assume that $\{\lambda_i^X\}$, $\{\lambda_j^Y\}$, $\{\mu_i\}$, $\{\nu_j\}$, and β are such that parameters from different sets are independent. For instance, μ_1 and υ_1 are independent, but υ_1 and υ_2 may be dependent. The term μ in the RC model is simply a normalizing constant, and it will not be included in our discussion. Since there is usually little interest in the $\{\lambda_i^X\}$ and $\{\lambda_j^Y\}$, we let each of $\{\lambda_i^X\}$ and each of $\{\lambda_j^Y\}$ have an improper uniform distribution over the entire real line.

One can incorporate a prior belief of likelihood-ratio dependence in the RC model by using prior distributions for the $\{\mu_i\}$ and $\{\upsilon_j\}$ that satisfy (1.2). One possibility is to let each set have the distribution of order statistics from some distribution. We shall let $(\mu_2,\cdots,\mu_{r-1})$ be order statistics of size $r-2$ from a uniform $U[1,r]$ distribution, and let $(\upsilon_2,\cdots,\upsilon_{c-1})$ be order statistics from a $U[1,c]$ distribution (recall that the model is parameterized so that $\mu_1 = \upsilon_1 = 1$, $\mu_r = r$, $\upsilon_c = c$). The uniform prior structure is useful, since it incorporates the category orderings and it also implies that

$$(3.1) \qquad E\mu_i = i, \quad E\upsilon_j = j, \quad \text{and} \quad E \log \theta_{ij} = E\beta.$$

Hence, when we take the scores to be order statistics from uniform distributions, the means of the local log odds ratios are constant; that is, the mean of the prior distribution corresponds to parameter values for the uniform association model.

Finally, one can select a prior distribution for β that reflects beliefs about the average of the local log odds ratios. Since β can take any value on the real line, we will use a member $N(\mu_\beta,\sigma_\beta)$ of the family of normal distributions. We would choose σ_β to be relatively larger when we are less sure of the strength of the association.

Now, let

$$\underset{\sim}{\theta}' = (\lambda_1^X, \cdots, \lambda_{r-1}^X, \lambda_1^Y, \cdots, \lambda_{c-1}^Y, \mu_2, \cdots, \mu_{r-1}, \nu_2, \cdots, \nu_{c-1}, \beta),$$

and again assume that $p(\underset{\sim}{x}|\underset{\sim}{\theta})$ is multinomial. The prior distribution just discussed is

$$(3.2) \qquad p(\underset{\sim}{\theta}|\mu_\beta, \sigma_\beta) \propto \frac{1}{\sqrt{2\pi}\,\sigma_\beta} e^{-(\beta-\mu_\beta)^2/2\sigma_\beta^2}, \quad \begin{matrix} 1 = \mu_1 \le \cdots \le \mu_r = r \\ 1 = \nu_1 \le \cdots \le \nu_c = c \end{matrix}.$$

The posterior distribution of $\underset{\sim}{\theta}$ is proportional to $p(\underset{\sim}{x}|\underset{\sim}{\theta})p(\underset{\sim}{\theta}|\mu_\beta, \sigma_\beta)$, or

$$(3.3) \qquad p(\underset{\sim}{\theta}|\underset{\sim}{x}, \mu_\beta, \sigma_\beta) \propto (\Pi\Pi\, m_{ij}^{x_{ij}}) e^{-(\beta-\mu_\beta)^2/2\sigma_\beta^2}, \quad \begin{matrix} 1 = \mu_1 \le \cdots \le \mu_r = r \\ 1 = \nu_1 \le \cdots \le \nu_c = c \end{matrix}.$$

The Bayes estimate of $\underset{\sim}{\theta}$ with respect to squared-error loss is the posterior mean of $\underset{\sim}{\theta}$ given $\underset{\sim}{x}$. A much simpler estimate to calculate (suggested by Lindley and Smith, 1972) is the posterior mode $\hat{\underset{\sim}{\theta}}$, which Leonard (1975) refers to as the generalized maximum likelihood estimate (GMLE). If the true scores are strictly ordered, then for large samples the posterior mode and the posterior mean are very similar, since the former estimate behaves like the ML estimate, which has an approximately normal distribution. The posterior mode may be considerably different from the posterior mean otherwise, since the posterior distribution of $\underset{\sim}{\theta}$ is then a severe truncation of an asymptotically normal distribution. Computation of this estimate will be discussed later in the section. Suppose that $\{\hat{\mu}_i\}$ and $\{\hat{\nu}_j\}$ in $\hat{\underset{\sim}{\theta}}$ are such that

$$\begin{aligned}
&\hat{\mu}_1 = \cdots = \hat{\mu}_{r_1} < \hat{\mu}_{r_1+1} = \cdots = \hat{\mu}_{r_2} < \cdots < \hat{\mu}_{r_{a-1}+1} = \cdots = \hat{\mu}_{r_a} \\
&\hat{\upsilon}_1 = \cdots = \hat{\upsilon}_{c_1} < \hat{\upsilon}_{c_1+1} = \cdots = \hat{\upsilon}_{c_2} < \cdots < \hat{\upsilon}_{c_{b-1}+1} = \cdots = \hat{\upsilon}_{c_b}.
\end{aligned} \tag{3.4}$$

It is shown in the Appendix that the GMLE satisfies the "generalized likelihood equations"

$$\hat{m}_{i+} = x_{i+}, \quad i = 1,\cdots,r, \tag{3.5a}$$

$$\hat{m}_{+j} = x_{+j}, \quad j = 1,\cdots,c, \tag{3.5b}$$

$$\sum_{R_1}\sum_{j} \hat{\upsilon}_j \hat{m}_{ij} = \sum_{R_1}\sum_{j} \hat{\upsilon}_j x_{ij} + (\hat{\beta}-\mu_\beta)/(r-1)\sigma_\beta^2. \tag{3.6a}$$

$$\sum_{R_k}\sum_{j} \hat{\upsilon}_j \hat{m}_{ij} = \sum_{R_k}\sum_{j} \hat{\upsilon}_j x_{ij}, \quad k = 2,\cdots,a-1, \tag{3.6b}$$

$$\sum_{R_a}\sum_{j} \hat{\upsilon}_j \hat{m}_{ij} = \sum_{R_a}\sum_{j} \hat{\upsilon}_j x_{ij} - (\hat{\beta}-\mu_\beta)/(r-1)\sigma_\beta^2. \tag{3.6c}$$

$$\sum_{i}\sum_{C_1} \hat{\mu}_i \hat{m}_{ij} = \sum_{i}\sum_{C_1} \hat{\mu}_i x_{ij} + (\hat{\beta}-\mu_\beta)/(c-1)\sigma_\beta^2. \tag{3.7a}$$

$$\sum_{i}\sum_{C_k} \hat{\mu}_i \hat{m}_{ij} = \sum_{i}\sum_{C_k} \hat{\mu}_i x_{ij}, \quad k = 2,\cdots,b-1 \tag{3.7b}$$

$$\sum_{i}\sum_{C_b} \hat{\mu}_i \hat{m}_{ij} = \sum_{i}\sum_{C_b} \hat{\mu}_i x_{ij} - (\hat{\beta}-\mu_\beta)/(c-1)\sigma_\beta^2 \tag{3.7c}$$

where $R_k = \{r_{k-1}+1,\cdots,r_k\}$ and $C_k = \{c_{k-1}+1,\cdots,c_k\}$, with $r_0 = c_0 = 0$.

Some interesting interpretations for this Bayes solution follow directly from these generalized likelihood equations. Equations (3.6) or (3.7) imply that

$$\sum\sum \hat{\mu}_i \hat{\nu}_j \hat{\pi}_{ij} = \sum\sum \hat{\mu}_i \hat{\nu}_j p_{ij} - (\hat{\beta}-\mu_\beta)/n\sigma_\beta^2, \tag{3.8}$$

where $\{p_{ij} = x_{ij}/n\}$. The observed correlation of the $\{\hat{\mu}_i\}$ and $\{\hat{\nu}_j\}$ is

$$\frac{\sum\sum \hat{\mu}_i \hat{\nu}_j p_{ij} - \left[\sum \hat{\mu}_i p_{i+}\right]\left[\sum \hat{\nu}_j p_{+j}\right]}{\sqrt{\left[\sum \hat{\mu}_i^2 p_{i+} - \left[\sum \hat{\mu}_i p_{i+}\right]\right]^2 \left[\sum \hat{\nu}_j^2 p_{+j} - \left[\sum \hat{\nu}_j p_{+j}\right]^2\right]}}.$$

The marginal observed and fitted distributions are identical, by (3.5), so equation (3.8) implies that the fitted correlation is greater than the observed correlation if the prior mean μ_β for β is relatively large (precisely, if μ_β exceeds the posterior estimate $\hat{\beta}$), whereas the fitted correlation is less than the observed correlation if the prior mean is relatively small $(\mu_\beta < \hat{\beta})$. Hence, the prior information results in an adjustment to the correlation, where the adjustment diminishes as $\hat{\beta}$ is closer to μ_β, for fixed σ_β and n. By comparison, the observed and fitted correlations are identical for the regular ML solution. As $\sigma_\beta \downarrow 0$ with μ_β and n fixed, it follows from equations (3.6)-(3.7) that $\hat{\beta} \longrightarrow \mu_\beta$. In other words, the more precise the prior information, the closer the fitted average local log odds ratio is forced to be to the prior mean μ_β.

As in the order-restricted ML approach for the RC model, it is nontrivial to give a routine that necessarily produces the partition corresponding to the GMLE. We have developed a program that incorporates the E04UAF subroutine from the NAG library (1984) for finding the point at which $p(\underset{\sim}{\theta}|\underset{\sim}{x},\mu_\beta,\sigma_\beta)$ is maximized.

4. ORDER-RESTRICTED BAYES SOLUTION FOR R OR C MODELS. A similar approach can be used for the R model or the C model. For the R model, for instance, the $\{\nu_j\}$ are fixed, monotone scores and the $\{\mu_i\}$ are parameters. We can incorporate the order restriction on the $\{\mu_i\}$ by letting the prior distribution of $(\mu_2,\cdots,\mu_{r-1})$ be that of order statistics of size $r-2$ from a uniform $U[1,r]$ distribution. We treat the other parameters the same way as in the RC model. Then,

$$E\mu_i = i \quad \text{and} \quad E \log \theta_{ij} = (\nu_{j+1}-\nu_j)E\beta, \tag{4.1}$$

so that for $\{\nu_j = j\}$ the mean of the prior distribution corresponds to parameter values for the uniform association model. The posterior distribution has form (3.3) for $1 = \mu_1 \le \cdots \le \mu_r = r$. The generalized likelihood equations are

$$\hat{m}_{i+} = x_{i+}, \qquad i = 1,\cdots,r, \tag{4.2a}$$

$$\hat{m}_{+j} = x_{+j}, \qquad j = 1,\cdots,c, \tag{4.2b}$$

$$\sum_{R_1}\sum_j \nu_j\hat{m}_{ij} = \sum_{R_1}\sum_j \nu_j x_{ij} + (\hat{\beta}-\mu_\beta)/(r-1)\sigma_\beta^2, \tag{4.3a}$$

$$\sum_{R_k}\sum_j \nu_j\hat{m}_{ij} = \sum_{R_k}\sum_j \nu_j x_{ij}, \qquad k = 2,\cdots,a-1, \tag{4.3b}$$

(4.3c) $$\sum_{R_a}\sum_j \upsilon_j\hat{m}_{ij} = \sum_{R_a}\sum_j \upsilon_j x_{ij} - (\hat{\beta}-\mu_\beta)/(r-1)\sigma_\beta^2.$$

From (4.2a) and (4.3b), the order-restricted solution equates the fitted row means $\left\{\sum_j \upsilon_j\hat{m}_{ij}/\hat{m}_{i+}\right\}$ with the observed row means when the rows in R_k are collapsed into a single row, $k = 2,\cdots,a-1$. A basic property of any solution satisfying the R model is that the $\{\hat{\beta}\hat{\mu}_i\}$ have the same ordering as these fitted row means. If $\hat{\beta} > 0$, for instance, then the fitted means for the order-restricted solution are monotone increasing. When $\hat{\beta} > \mu_\beta$ in that case, note from (4.3a) and (4.3c) that there is a shrinkage of the fitted means in the 1st and a^{th} sets of rows towards the grand mean for the table, the degree of shrinkage increasing as σ_β decreases. When $\hat{\beta} < \mu_\beta$, the reverse applies. In an important special case, if $\left\{\sum \upsilon_j x_{ij}/x_{i+}, i = 2,\cdots,r-1\right\}$ are monotone increasing, and if the (μ_β,σ_β) combination produces a $\hat{\beta}$ for which

$$\sum_j \upsilon_j x_{1j}/x_{1+} + (\hat{\beta}-\mu_\beta)/(r-1)x_{1+}\,\sigma_\beta^2 < \sum_j \upsilon_j x_{2j}/x_{2+}$$

and

$$\sum_j \upsilon_j x_{r-1,j}/x_{r-1,+} < \sum_j \upsilon_j x_{rj}/x_{r+} - (\hat{\beta}-\mu_\beta)/(r-1)x_{r+}\,\sigma_\beta^2,$$

it follows that $\hat{\mu}_1 < \hat{\mu}_2 <\cdots< \hat{\mu}_r$ for the Bayes solution (i.e., $R_k = k$, all k). In this sense, prior beliefs reflecting strong association (through a large μ_β) have the effect of moving the estimates away from the boundary of the restricted parameter space.

The generalized likelihood equations for the Bayes solutions are asymptotically equivalent to those for the order-restricted ML solution as $\sigma_\beta \longrightarrow \infty$ (for fixed n and μ_β) or as $n \longrightarrow \infty$ (for fixed μ_β and $\sigma_\beta > 0$). In other words, the Bayes estimates more closely resemble the ML estimates as the prior information about the strength of association is more vague, or as the sample size grows. Also, in the particular case that the Bayes estimate $\hat{\beta}$ equals the prior mean μ_β, the Bayes solution is identical to the order-restricted ML solution.

Of the parameters in $\underset{\sim}{\theta}$, estimation of β is usually paramount, since that parameter represents the strength of association (through the average of the local log odds ratios). In this regard it is useful to attach a standard error to any estimate of β. This is difficult to do with our Bayesian approach, since the posterior distribution of β requires numerical integration. If the parameter scores in model (1.1) are truly strictly ordered, then the Bayes and the order-restricted ML estimates of β have the same asymptotic distributions as the ordinary ML estimate of β. For large samples, therefore, it would usually suffice to use the asymptotic standard error of the ordinary ML estimate, which can be obtained from the inverse of the information matrix.

5. **EXAMPLE.** We will illustrate the Bayes and ML estimates for model (1.1) using the data in Table 1, taken from Srole et al. (1978, p. 289), which were analyzed in Goodman (1979). Theory and research in mental health studies have consistently suggested that mental health continually improves with increasing socioeconomic status (SES) of subjects or their parents (see, e.g., Dohrenwend and Dohrenwend, 1969). Hence, if model (1.1) holds, it is reasonable to assume that the score parameters are strictly monotonic, with $\beta > 0$. According to standard ML fitting, model (1,1) fits quite well, with $G^2 = 3.57$ and $df = 8$ for the general RC model. However, the ML estimates

Table 1. Cross-classification of subjects according to their mental health and parents' socioeconomic status*

Mental Health Status	Parents' Socioeconomic Status A	B	C	D	E	F
Well	64	57	57	72	36	21
Mild Symptom Formation	94	94	105	141	97	71
Moderate Symptom Formation	58	54	65	77	54	54
Impaired	46	40	60	94	78	71

*A = high; F = low.
Source: Goodman (1979)

Table 2. Estimated association parameters in RC model for Table 1.

Parameter		Specification of $N(\mu_\beta, \sigma_\beta)$ prior for β					
μ_β		.1	.2	.3	.1	.2	.3	0
σ_β		.1	.1	.1	.2	.2	.2	∞
	ML	Bayes Estimates						
β	.101	.101	.104	.108	.101	.102	.103	.101
μ_1	1.0	1.0	1.0	1.0	1.0	1.0	1.0	1.0
μ_2	2.49	2.49	2.51	2.52	2.49	2.50	2.50	2.49
μ_3	2.76	2.76	2.77	2.79	2.76	2.76	2.77	2.76
μ_4	4.0	4.0	4.0	4.0	4.0	4.0	4.0	4.0
ν_1	1.0	1.0	1.0	1.0	1.0	1.0	1.0	1.0
ν_2	.98	1.00	1.06	1.13	1.00	1.01	1.03	1.00
ν_3	2.26	2.27	2.29	2.31	2.27	2.27	2.28	2.27
ν_4	2.94	2.95	2.94	2.94	2.95	2.94	2.94	2.95
ν_5	4.62	4.62	4.56	4.51	4.62	4.61	4.59	4.62
ν_6	6.0	6.0	6.0	6.0	6.0	6.0	6.0	6.0
G^2	3.57	3.57	3.61	3.72	3.57	3.57	3.58	3.57

$\hat{\nu}_1$ and $\hat{\nu}_2$ for the RC model are slightly out of order, as seen in the first column of Table 2.

We analyzed these data with the Bayesian approach having normal prior distribution for β with $\mu_\beta = .2$ and $\sigma_\beta = .1$. We chose this prior because almost all of its support is on the positive part of the real line, with a distance of two standard deviations encompassing both independence ($\beta = 0$) and strong association ($\beta = .4$ corresponds to an odds ratio of $\exp[.4(4-1)(6-1))] = 403.4$ for the four corner cells). The GMLE of the association parameters in the posterior distribution is $\hat{\beta} = .104$, $\hat{\underset{\sim}{\mu}} = (1.0, 2.51, 2.77, 4.0)$, $\hat{\underset{\sim}{\nu}} = (1.0, 1.06, 2.29, 2.94, 4.56, 6.0)$. This fit gives expected frequencies having $G^2 = 3.61$. We feel that the Bayes estimates are more pleasing than the regular ML estimates for these data, since they result in a fit that is practically as good as the ML fit but which has uniformly positive local log odds ratios. That is, both sets of score parameters have strictly monotonic estimates, leading to simpler interpretation for the estimates and for the local log odds ratios.

From the likelihood and generalized likelihood equations, it follows that if $|\mu_\beta| > |\hat{\beta}|$ and σ_β is sufficiently small, the Bayes estimates of the score parameters may be strictly ordered, even if the ML estimates are not. Table 2 also contains the GMLE's for several choices of μ_β and σ_β in the Bayes approach, in order to illustrate the dependence on that choice. The estimates for $\sigma_\beta = \infty$ are the order-restricted ML estimates. For these data, the sample size is large, and widely disparate choices of prior parameters produce similar results.

6. **COMMENTS**. The analyses proposed in Sections 3 and 4 are merely intended to suggest how Bayesian methods can produce order-restricted inference for model (1.1). There are several ways in which these analyses can be amended to reflect somewhat different conditions. For instance:

6.1 *Alternative Prior Distributions for Score Parameters.* The assumption (through a prior distribution) that constraint (1.2) holds is quite a strong one. In some applications one would be less certain that the model satisfies likelihood-ratio dependence. Then, it could be more appropriate to use a prior for which the parameter scores are ordered in expectation, but need not be ordered with probability one. For instance, for the RC model one could assume that the $\{\mu_i\}$ are independent $\{N(u_i, \sigma_1)\}$ and that the $\{\nu_j\}$ are independent $\{N(v_j, \sigma_2)\}$, for values $1 = u_1 < u_2 < \cdots < u_r = r$, $1 = v_1 < v_2 < \cdots < v_c = c$ and σ_1 and σ_2 chosen by the researcher. For $\{u_i = i\}$ and $\{v_j = j\}$, again $E\mu_i = i$, $E\nu_j = j$, and $E \log \theta_{ij} = E\beta$, corresponding to uniform association. Since $P(\mu_{i+1} > \mu_i) = \Phi(1/\sqrt{2}\sigma_1)$ and $P(\nu_{j+1} > \nu_j) = \Phi(1/\sqrt{2}\sigma_2)$, greater prior confidence that the scores are monotone is reflected by taking smaller values for σ_1 and σ_2. This prior structure has the effect of moving the ML estimates towards ordered scores. The posterior probability that the Bayes estimates of the scores are ordered increases as σ_1 and σ_2 decrease. The limiting case $\{\{u_i = i\}, \{v_j = j\}, \sigma_1 = \sigma_2 = 0, \sigma_\beta = \infty\}$ corresponds to the ML fit of the uniform association model, the limiting case $\{\sigma_1 = \infty, \sigma_2 = 0, \sigma_\beta = \infty\}$ corresponds to the ML fit of the row effects model, the limiting case $\{\sigma_1 = 0, \sigma_2 = \infty, \sigma_\beta = \infty\}$ corresponds to the ML fit of the column effects model, and the limiting case $(\sigma_1 = \sigma_2 = \sigma_\beta = \infty\}$ corresponds to the ML fit of the RC model.

6.2 ***Empirical Bayes Estimation of Model Parameters.*** One must specify μ_β and σ_β (in the normal prior distribution for β) to use the Bayes approach described in Sections 3 and 4. Alternatively, one could use a parametric empirical Bayes approach whereby μ_β and

σ_β are data dependent. For instance, we could consider the influence of $(\mu_\beta, \sigma_\beta)$ in the marginal distribution of $\underset{\sim}{X}$,

$$m(\underset{\sim}{x};\mu_\beta,\sigma_\beta) = \int p(\underset{\sim}{x}|\underset{\sim}{\theta})p(\underset{\sim}{\theta}|\mu_\beta,\sigma_\beta)d\underset{\sim}{\theta}.$$

Given $\underset{\sim}{x}$, one could treat $m(\underset{\sim}{x};\mu_\beta,\sigma_\beta)$ as a pseudo likelihood, and then use the (μ_β,σ_β) combination that maximizes m to specify the prior $p(\underset{\sim}{\theta}|\mu_\beta,\sigma_\beta)$. Unfortunately, straightforward maximization of m seems to be intractable. In somewhat related work, Laird (1978) and Chuang (1982) used the EM algorithm to find an estimate for the variance σ^2 in a prior distribution by maximizing the marginal density of $\underset{\sim}{X}$ given σ. Their approach uses the fact that the distribution of $\underset{\sim}{Y} = (\underset{\sim}{X},\underset{\sim}{\theta})$ with $\underset{\sim}{\theta}$ viewed as missing is in the exponential family, and then equations to find the mode for exponential family distributions are readily available. An approximation is used in the E step by both authors. However, the ordering constraints on the score parameters imply that the distribution of $(\underset{\sim}{Y}|\mu_\beta,\sigma_\beta)$ is no longer in the exponential family, and the approximation adopted by Laird and Chuang is inappropriate.

Other approaches could be explored to obtain data-determined (μ_β,σ_β) for use in our Bayesian fit of the RC model. For instance, a simple approach is to use

$$\hat{\mu}_\beta = \sum\sum \log \hat{\theta}_{ij}/(r-1)(c-1), \quad \hat{\sigma}^2_\beta = \frac{\sum\sum(\log \hat{\theta}_{ij} - \hat{\mu}_\beta)^2}{(rc-r-c)(r-1)(c-1)}$$

where $\{\log \hat{\theta}_{ij}\}$ are estimates of the local log odds ratios. Here $\hat{\sigma}^2_\beta$ mimics the form of the variance for β if it were regarded as an average of (r-1)(c-1) uncorrelated local association components. This value decreases as the data more closely follow the U model, which is the model corresponding to parameter values that are the means of the regular Bayes prior. Unless the U model fits perfectly in the population, $\hat{\sigma}^2_\beta$ converges in probability to a positive constant as $n \longrightarrow \infty$, and this empirical Bayes GMLE is asymptotically equivalent to the order-restricted ML estimate.

6.3 *Tables having ordinal and nominal classifications.* The methods in this paper can be generalized for the analysis of multidimensional tables or tables that contain nominal as well as ordinal variables. To illustrate, suppose that the row variable is nominal and the column variable is ordinal. Then, in model (1.1), it is inappropriate to impose an ordering on the $\{\mu_i\}$, and their prior distributions should take this into account.

APPENDIX.

Derivation of generalized likelihood equations

In this appendix we derive the "generalized likelihood equations" satisfied by the posterior mode for the Bayesian analysis presented in Section 3.

Finding the mode of the posterior distribution is equivalent to maximizing the right-hand side of equation (3.3), with the RC model substituted for m_{ij}. Differentiating

$$L = \log p(\underset{\sim}{\theta} | \underset{\sim}{x}, \mu_\beta, \sigma_\beta) - \tau(\sum \sum m_{ij} - n)$$

separately with respect to λ_i^X and λ_j^Y and setting the results equal to zero produces (3.5a) and (3.5b), with the Lagrange multiplier $\tau = 1$.

Consider the maximization of L with respect to the row scores $\{\mu_i\}$. It is convenient here to use a reparameterized version of the model in which $\beta\mu_i$ is replaced by $\beta + \sum_{k=1}^{i-1} \Delta_k$, $i = 2,\cdots,r$. Then $\mu_r = r$ implies that $\beta = \sum_{k=1}^{r-1} \Delta_k/(r-1) = \bar{\Delta}$, so that $\mu_{i+1}-\mu_i = \Delta_i/\bar{\Delta}$, $i = 1,\cdots,r-1$. The constraints on $\{\mu_i\}$ correspond to constraints on $\{\Delta_k\}$ of $\{\text{all } \Delta_k \geq 0\}$ or $\{\text{all } \Delta_k \leq 0\}$.

Let $\{\hat{\mu}_i\}$ and $\{\hat{\nu}_j\}$ denote the GMLE's that satisfy the given order restrictions. Suppose there are a distinct $\hat{\mu}_i$ values $\hat{\mu}_{(1)} < \hat{\mu}_{(2)} <\cdots< \hat{\mu}_{(a)}$. Partition the rows $\{1,2,\cdots,r\}$ into $\{R_1,\cdots,R_a\}$, where R_k contains consecutive integers such that $\hat{\mu}_i = \hat{\mu}_{(k)}$ if $i \in R_k$. Suppose that the number of elements in R_1 equals r_1; that is, $\hat{\mu}_1 =\cdots= \hat{\mu}_{r_1} < \hat{\mu}_{r_1+1}$, so that $\hat{\Delta}_1 =\cdots= \hat{\Delta}_{r_1-1} = 0$ and $\hat{\Delta}_{r_1} \neq 0$. With respect to the $\{\Delta_k\}$,

$$L = \text{const.} + \sum_i \sum_j x_{ij}\left(\bar{\Delta} + \sum_{k=1}^{i-1} \Delta_k\right)\nu_j - (\bar{\Delta}-\mu_\beta)^2/2\sigma_\beta^2$$

$$- \left\{\sum_i \sum_j \exp\left[\mu + \lambda_i^X + \lambda_j^Y + \left[\bar{\Delta} + \sum_{k=1}^{i-1} \Delta_k\right]\nu_j\right] - n\right\}.$$

Thus

$$\frac{\partial L}{\partial \Delta_{r_1}} = \sum_{i=r_1+1}^{r} \sum_j x_{ij}\nu_j - (\bar{\Delta}-\mu_\beta)/(r-1)\sigma_\beta^2$$

$$- \sum_{i=r_1+1}^{r} \sum_j \nu_j \exp\left[\mu + \lambda_i^X + \lambda_j^Y + \left[\bar{\Delta} + \sum_{k=1}^{i-1} \Delta_k\right]\nu_j\right]$$

$$+ \sum_i \sum_j \nu_j x_{ij}/(r-1) - \left\{\sum_i \sum_j \nu_j \exp\left[\mu + \lambda_i^X + \lambda_j^Y + \left[\bar{\Delta} + \sum_{k=1}^{i-1} \Delta_k\right]\nu_j\right]\right\}/(r-1),$$

which is a strictly monotone decreasing function of Δ_{r_1}. In particular, $\partial^2 L/\partial \Delta_{r_1}^2 < 0$, so that L is a strictly concave function of Δ_{r_1}. Since $|\hat{\Delta}_{r_1}| > 0$ is in the interior of $[0,\infty)$, it follows that $\partial L/\partial \Delta_{r_1} = 0$ at the GMLE solution. Thus,

$$\sum_{i=r_1+1}^{r} \sum_j \hat{m}_{ij}\hat{\nu}_j = \sum_{i=r_1+1}^{r} \sum_j x_{ij}\hat{\nu}_j - (\hat{\beta}-\mu_\beta)/(r-1)\sigma_\beta^2,$$

where the terms in the last row for $\partial L/\partial \Delta_{r_1}$ vanished because $\{\hat{m}_{+j} = x_{+j}\}$ at the GMLE. Next, if $\hat{\mu}_{r_1+1} = \cdots = \hat{\mu}_{r_2} < \hat{\mu}_{r_2+1}$, so that $\hat{\Delta}_{r_2} \neq 0$, we obtain

$$\sum_{i=r_2+1}^{r} \sum_j \hat{m}_{ij}\hat{\nu}_j = \sum_{i=r_2+1}^{r} \sum_j x_{ij}\hat{\nu}_j - (\hat{\beta}-\mu_\beta)/(r-1)\sigma_\beta^2.$$

This argument can be repeatedly applied until, finally, from $\hat{\mu}_{r_{a-1}} < \hat{\mu}_{r_{a-1}+1} = \cdots = \hat{\mu}_r$, so $\hat{\Delta}_{r_{a-1}} \neq 0$, we obtain

$$\sum_{i=r_{a-1}+1}^{r} \sum_j \hat{m}_{ij}\hat{\nu}_j = \sum_{i=r_{a-1}+1}^{r} \sum_j x_{ij}\hat{\nu}_j - (\hat{\beta}-\mu_\beta)/(r-1)\sigma_\beta^2.$$

Sequential subtraction of these equations gives (3.6). The analogous argument applied to the $\{\nu_j\}$ yields (3.7).

REFERENCES

Agresti, A. (1984). *Analysis of Ordinal Categorical Data*, New York: Wiley.

Agresti, A., Chuang, C. & Kezouh, A. (1986). Order-Restricted Score Parameters in Association Models for Contingency Tables. Submitted for publication.

Chuang, C. (1982). Empirical Bayes Methods for a Two-Way Multiplicative-Interaction Model. *Comm. Statist.* **A11**, 2977-2989.

Dixon, W.J. (1979). *BMDP Statistical Software*, Los Angeles, CA: University of California Press.

Dohrenwend, B.P. & Dohrenwend, B.S. (1969). *Social Status and Psychological Disorder: A Causal Inquiry*, New York: Wiley.

Gill, P.E., Murray, W. & Wright, M.H. (1981). *Practical Optimization*, New York: Academic Press.

Goodman, L.A. (1979). Simple Models for the Analysis of Association in Cross-Classifications Having Ordered Categories, *J. Amer. Statist. Assoc.*, **74**, 537-552.

Goodman, L.A. (1981a). Association Models and Canonical Correlation in the Analysis of Cross-Classifications Having Ordered Categories, *J. Amer. Statist. Assoc.* **76**, 320-334.

Goodman, L.A. (1981b). Association Models and the Bivariate Normal Distribution in the Analysis of Cross-Classifications Having Ordered Categories, *Biometrika* **68**, 347-355.

Laird, N.M. (1978). Empirical Bayes Methods for Two-Way Contingency Tables, *Biometrika* **65**, 581-590.

Lehmann, E.L. (1966). Some Concepts of Dependence, *Ann. Math. Statist.* **37**, 1137-1153.

Leonard, T. (1975). Bayesian Estimation Methods for Two-Way Contingency Tables, *J. Roy. Statist. Soc.* **B37**, 23-37.

Lindley, D.V. & Smith, A.F.M. (1972). Bayes Estimates for the Linear Model, *J. Roy. Statist. Soc.* **B34**, 1-18.

NAG Fortran Mini-Manual, Mark 11 (1984), Numerical Algorithms Group Inc., Downers Grove, IL.

Srole, L., Langner, T.S., Michael, S.T., Kirkpatrick, P., Opler, M.K. & Rennie, T.A.C. (1978). *Mental Health in the Metropolis: The Midtown Manhattan Study*, Rev. ed., New York: NYU Press.

Alan Agresti
Department of Statistics
University of Florida
Gainesville, FL 32611

Christy Chuang
7293-32-2
The Upjohn Company
Kalamazoo, MI 49001

A METHOD FOR FINDING PROJECTIONS ONTO THE INTERSECTION OF CONVEX SETS IN HILBERT SPACES[1]

James P. Boyle
Richard L. Dykstra
The University of Iowa

AMS 1980 subject classifications: Primary 49D99, 65D99.
Key words and phrases: projections, least squares, regression, convex constraints, iterative methods, Hilbert space, weak convergence, strong convergence, constrained optimization.

ABSTRACT

Many problems require the ability to find least squares projections onto convex regions. Here it is shown that if the constraint region can be expressed as a finite intersection of simpler convex regions, then one can obtain the projection onto the intersection by performing a series of projections only onto the simpler regions. This may facilitate finding least squares projections in more complicated spaces since the approach is valid for a general Hilbert space rather than just R^n.

1. **INTRODUCTION.** Estimating parameters by minimizing a sum of squares has a long history dating as far back as Gauss. Probably the most popular application of this least squares method is the estimation of β in the linear model $Y = X\beta+\varepsilon$, where β is either unrestricted or constrained to satisfy restrictions of the form $A\beta = r$

---

1. This research was supported in part by ONR Contract N00014-83-K-0249.

or $A\beta \leq r$. This least squares estimate $\hat{\beta} \in \mathbb{R}^k$ in the unrestricted case is such that $X\hat{\beta}$ is the projection, with respect to the usual Euclidean metric, of $Y \in \mathbb{R}^n$ onto the subspace generated by the columns of X. In the restricted cases the estimate $\tilde{\beta} \in \mathbb{R}^k$ is simply the projection of $\hat{\beta}$ onto a closed convex set, where now the metric is determined by the inner product $\beta_1 \cdot \beta_2 = \beta_1' X' X \beta_2$. Moreover, these estimates are the maximum likelihood estimates if ϵ is $N(0, \sigma^2 I)$.

Other situations where a projection solves an estimation problem abound. For example, Barlow et al. (1972) and Barlow and Brunk (1972) show that many restricted maximum likelihood estimates are either least squares projections or simple transforms of least squares projections in finite-dimensional Euclidean space. Also, in what may be an infinite-dimensional setting, it is well-known that the conditional expectation of a square-integrable random variable Y given a sub σ-field D is the projection of Y in an L^2 space onto the subspace of D-measurable square-integrable random variables. The growing area of smoothing splines involves L^2 type projections in various settings.

Thus, we see that many interesting problems are of the form

$$\underset{f \in C}{\text{Minimize}} \; \|g - f\|, \tag{1.1}$$

where g is some fixed element of an inner product space and C is a closed convex set.

It is the purpose of this paper to generalize a result due to Dykstra (1983) who develops an algorithm for projecting an element in a finite-dimensional inner product space onto a closed convex cone K when K can be written as $K_1 \cap \cdots \cap K_r$, and each K_i is also a closed convex cone. Dykstra's idea is that often it is easy to project onto the individual K_i's, and this fact should be helpful in finding the solution to the more complicated problem of projecting onto K. In particular, his algorithm can be employed in the least squares estimation of β in the linear model subject to constraints of the form $A\beta \leq 0$, where $K_i = \{\beta \mid a_i'\beta \leq 0\}$. Here a_i' is the ith row of the $r \times k$ matrix A. This paper proves that his algorithm actually converges correctly in an infinite-dimensional Hilbert space setting even when the K_i's are replaced by arbitrary closed convex sets C_i. It is of interest to note that Von Neumann (1950) has proven that successive cyclic projections onto the C_i's, where the C_i's are closed subspaces, converge in a Hilbert space setting to the projection onto the intersection. This was also shown independently by Wiener (1955). Dykstra's procedure reduces to this procedure when the C_i's are closed subspaces.

In Section 2 the specific Hilbert space setting is detailed along with a description of the algorithm. The important concept of weak convergence, which plays a fundamental role in infinite-dimensional spaces, is discussed. This section also develops standard notation. Section 3 presents a proof of convergence, and Section 4 discusses a few applications. Section 5 offers some concluding remarks.

2. BACKGROUND AND A DESCRIPTION OF THE ALGORITHM. Let H be any real Hilbert space with inner product $x \cdot y$ defined for all $x, y \in H$. Recall that the inner product induces a complete norm $\|\cdot\|$ on H defined by $\|x\| = (x \cdot x)^{1/2}$. A sequence $\{x_n\} \in H$ is said to converge strongly to $x \in H$ if $\|x_n - x\| \longrightarrow 0$. We denote this by

$x_n \xrightarrow{s} x$. We say $\{x_n\}$ converges weakly to x if $x_n \cdot y \longrightarrow x \cdot y$ for all $y \in H$. This weak convergence is denoted by $x_n \xrightarrow{w} x$, and it is trivial to prove that strong convergence always implies weak convergence. The reverse implication is only true when H is finite-dimensional and thus weak convergence is a different concept in infinite-dimensional spaces.

An important well-known property of finite-dimensional inner product spaces is the Bolzano-Wierstrass Theorem. Specifically, any bounded sequence admits a strongly convergent subsequence. This does not hold in infinite-dimensional Hilbert spaces. However, the important weak compactness property does hold, i.e., any bounded sequence admits a weakly convergent subsequence. (See, for example, Balakrishnan (1971), p. 15.)

We also cite the following standard theorem (see Luenberger (1969), p.69).

THEOREM 1. *Let* g *be any element of* H *and let* $C \subseteq H$ *be any nonempty closed convex set. Then there is a unique* $g^* \in C$ *which solves* (1.1). *This minimizing element* g^* *is completely characterized by the condition*

$$(g-g^*) \cdot (g^*-f) \geq 0, \quad \forall f \in C. \tag{2.1}$$

When $C = C_1 \cap \cdots \cap C_r$, where each C_i is closed and convex, we propose an algorithm for the solution of Problem (1.1) which requires only the ability to find projections onto the C_i. We begin with the first cycle.

Cycle 1:

1) Project g onto C_1 and obtain $g_{11} = g+I_{11}$.

2) Project g_{11} onto C_2 to obtain $g_{12} = g_{11}+I_{12}$
$= g+I_{11}+I_{12}$.
$\vdots$

r) Project $g_{1,r-1}$ onto C_r to obtain $g_{1r} = g_{1,r-1}+I_{1r}$
$= g+I_{11}+I_{12}+\cdots+I_{1r}$.

After the first cycle, instead of projecting g_{1r} onto C_1, we first remove the initial increment I_{11} and then project. Specifically, the steps for the second cycle proceed as follows:

Cycle 2:

1) Project $g_{1r}-I_{11}$ onto C_1 to obtain $g_{21} = g_{1r}-I_{11}+ I_{21}$
$= g+I_{21}+I_{12}+\cdots+I_{1r}$.

2) Project $g_{21}-I_{12}$ onto C_2 and obtain $g_{22} = g_{21}-I_{12} +I_{22}$
$= g+I_{21}+I_{22}+I_{13}+\cdots+I_{1r}$.
$\vdots$

r) Project $g_{2,r-1}-I_{1r}$ onto C_r yielding
$g_{2r} = g_{2,r-1}-I_{1r}+I_{2r} = g+I_{21}+I_{22}+\cdots+I_{2r}$.

Continuing this routine, that of removing the increment in the previous cycle associated with C_i before projecting onto C_i, generates the infinite arrays $\{g_{ni}\}$ and $\{I_{ni}\}$, where $n \geq 1$ and $1 \leq i \leq r$. Note that the following relations are valid for $n \geq 1$ and $i = 2,3,\cdots,r$.

(2.2)
$$\text{a)}\quad g_{n-1,r}-g_{n1} = I_{n-1,1}-I_{n1}$$
$$\text{b)}\quad g_{n,i-1}-g_{ni} = I_{n-1,i}-I_{ni}$$

where for convenience we set $g_{0r} = g$ and $I_{0i} = 0$ for all i. We note the characterization

$$g_{ni} = g+I_{n1}+\cdots+I_{ni}+I_{n-1,i+1}+\cdots+I_{n-1,r}, \tag{2.3}$$

where again $n \geq 1$ and $1 \leq i \leq r$.

We state the following theorem.

THEOREM 2. *For any* $1 \leq i \leq r$ *the sequence* $\{g_{ni}\}$ *converges strongly to* g^*, *i.e.*, $\|g_{ni}-g^*\| \longrightarrow 0$ *as* $n \to +\infty$.

Before proceeding with the proof of Theorem 2, we borrow a lemma from Dykstra (1983).

LEMMA 3. *Let* $\{a_n\}$ *be a sequence of nonnegative real numbers with* $\sum_{m=1}^{+\infty} a_n^2 < +\infty$. *Then there exists a subsequence* $\{a_{n_j}\}$ *such that*

$$\sum_{n=1}^{n_j} a_m a_{n_j} \longrightarrow 0 \quad as \quad j \to +\infty.$$

3. **PROOF OF THEOREM 2.** Consider the equalities $\|g-g^*\|^2 = \|g_{11}-g^*-I_{11}\|^2 = \|g_{11}-g^*\|^2 + 2(g-g_{11})\cdot(g_{11}-g^*) + \|I_{11}\|^2$, where the middle term in the last expression is nonnegative by (2.1). That is, g_{11} is the projection of g onto C_1 and $g^* \in C_1$. We do a similar decomposition on $\|g_{11}-g^*\|$ and conclude $\|g-g^*\|^2 = \|g_{12}-g^*\|^2 + 2(g-g_{11}) \cdot (g_{11}-g^*) + 2(g_{11}-g_{12}) \cdot (g_{12}-g^*) + \|I_{11}\|^2 + \|I_{12}\|^2$, where again all terms are nonnegative. This process can be continued through the first cycle obtaining

(3.1) $$\|g-g^*\|^2 = \|g_{1r}-g^*\|^2 + 2\sum_{i=1}^{r}(g_{1,i-1}-g_{1i})\cdot(g_{1i}-g^*) + \sum_{i=1}^{r}\|I_{1i}\|^2,$$

(where $g_{10} = g$) with all terms being nonnegative. Now, since $\|g_{1r} - g^*\|^2 = \|g_{21} - g^* + I_{11} - I_{21}\|^2 = \|g_{21}-g^*\|^2 - 2(g-g_{11})\cdot(g_{21}-g^*) + 2(g_{1r}-I_{11}-g_{21})\cdot(g_{21}-g^*) + \|I_{11}-I_{21}\|^2$, we may substitute in (3.1) and obtain

$$\|g-g^*\|^2 = \|g_{21}-g^*\|^2 + 2(g-g_{11})\cdot(g_{11}-g_{21})$$
$$+2(g_{1r}-I_{11}-g_{21})\cdot(g_{21}-g^*) + 2\sum_{i=2}^{r}(g_{1,i-1}-g_{1i})\cdot(g_{1i}-g^*)$$
$$+\sum_{i=1}^{r}\|I_{1i}\|^2 + \|I_{11}-I_{21}\|^2.$$

Once again all terms are nonnegative. In general, this decomposition process can be continued into the nth cycle to obtain $\|g-g^*\|^2 = \|g_{ni}-g^*\|^2$ + (sum of nonnegative terms). For $i = r$ we have

$$\|g-g^*\|^2 = \|g_{nr}-g^*\|^2 + \sum_{m=1}^{n}\sum_{i=1}^{r}\|I_{m-1,i}-I_{mi}\|^2$$
(3.2) $$+ 2\sum_{m=1}^{n-1}\sum_{i=1}^{r}(g_{m,i-1}-I_{m-1,i}-g_{mi})\cdot(g_{mi}-g_{m+1,i})$$
$$+ 2\sum_{i=1}^{r}(g_{n,i-1}-I_{n-1,i}-g_{ni})\cdot(g_{ni}-g^*),$$

where we set $g_{m0} = g_{m-1,r}$ for any m and, of course, $I_{0i} = 0$ for all i.

Because all terms in (3.2) are nonnegative for all n, we

clearly have the infinite sum

$$\sum_{m=1}^{+\infty} \sum_{i=1}^{r} \|I_{m-1,i}-I_{mi}\|^2 < +\infty. \tag{3.3}$$

But (3.3), along with a) and b) of (2.2), implies that the sequence of successive increments $\|g_{11}-g_{12}\|, \|g_{12}-g_{13}\|, \cdots, \|g_{1r}-g_{21}\|, \cdots, \|g_{n,i-1}-g_{ni}\|$ tends to zero. Hence, $g_{nr} \xrightarrow{s} g^*$ if and only if $g_{ni} \xrightarrow{s} g^*$ for all i, and it is sufficient to examine the sequence $\{g_{nr}\}_{n=1}^{+\infty}$.

Now, let f be any element of $C = \bigcap_{i=1}^{r} C_i$. For $n \geq 1$ we can write

$$\begin{aligned}
(g_{nr}-g)\cdot(g_{n1}-f) &= (I_{n1}+\cdots+I_{nr})\cdot(g_{n1}-f) \\
&= I_{n1}\cdot(g_{n1}-f)+\cdots+I_{nr}\cdot(g_{n1}-f) \\
&= I_{n1}\cdot(g_{n1}-g_{n1}+g_{n1}-f) + I_{n2}\cdot(g_{n1}-g_{n2}+g_{n2}-f) + \\
&\qquad \cdots + I_{nr}\cdot(g_{n1}-g_{nr}+g_{nr}-f),
\end{aligned}$$

which equals

$$\begin{aligned}
&I_{n2}\cdot(g_{n1}-g_{n2}) + \cdots + I_{nr}\cdot(g_{n1}-g_{nr}) \\
&\qquad + \{I_{n1}\cdot(g_{n1}-f) + \cdots + I_{nr}\cdot(g_{nr}-f)\}.
\end{aligned} \tag{3.4}$$

Note that every term in the brackets is nonpositive by the criterion (2.1). Now

$$(3.5)\qquad |I_{n2}\cdot(g_{n1}-g_{n2}) + \cdots + I_{nr}\cdot(g_{n1}-g_{nr})|$$
$$\le \|I_{n2}\|\|g_{n1}-g_{n2}\| + \cdots + \|I_{nr}\|\|g_{n1}-g_{nr}\|,$$

by properties of absolute value and the Cauchy-Schwarz inequality. Moreover, $\|g_{n1}-g_{ni}\| = \|(g_{n1}-g_{n2}) + (g_{n2}-g_{n3}) + \cdots + (g_{n,i-1}\cdot g_{ni})\|$ $\le \|g_{n1}-g_{n2}\| + \cdots + \|g_{n,r-1}-g_{nr}\|$. If we denote this last sum by a_n, then (3.5) is no greater than $\sum_{i=2}^{r} \|I_{ni}\|a_n$. We also have

$$\|I_{ni}\| = \|\sum_{m=1}^{n} (I_{mi}-I_{m-1,i})\| \le \sum_{m=1}^{n} \|I_{mi}-I_{m-1,i}\|.$$

Hence, (3.5) is bounded above by

$$\sum_{m=1}^{n} \sum_{i=2}^{r} \|I_{mi}-I_{m-1,i}\|a_n = \sum_{m=1}^{n} \sum_{i=2}^{r} \|g_{m,i-1}-g_{mi}\|a_n = \sum_{m=1}^{n} a_m a_n,$$

by (2.2) and the definition of a_n. Set $t_i = \|I_{n-1,i}-I_{ni}\|$ for $i = 2,\cdots,r$. We have $a_n^2 = \left[\sum_{i=2}^{r} t_i\right]^2 = \sum_{i=2}^{r} t_i^2 + 2\sum_{i<j} t_i t_j$, where there are $(r-1)(r-2) \div 2$ terms in the last sum. Also, since $(t_i-t_j)^2 \ge 0$, we see that $2t_i t_j \le t_i^2+t_j^2 \le \sum_{i=2}^{r} t_i^2$. Thus,

$$a_n^2 \le \left[\frac{(r-1)(r-2)}{2} + 1\right] \sum_{i=2}^{r} \|I_{n-1,i}-I_{ni}\|^2,$$

and we conclude that

$$\sum_{n=1}^{+\infty} a_n^2 \le \left[\frac{(r-1)(r-2)}{2} + 1\right] \sum_{n=1}^{+\infty} \sum_{i=2}^{r} \|I_{n-1,i}-I_{ni}\|^2.$$

But the right side of the above is finite by (3.3) so that $\sum_{n=1}^{+\infty} a_n^2 < +\infty$.

We are now able to apply Lemma 3, obtaining a subsequence $\{a_{n_j}\}$ such that $\sum_{m=1}^{n_j} a_m a_{n_j} \longrightarrow 0$ as $j \longrightarrow +\infty$. But we have shown that (3.5) is bounded above by $\sum_{m=1}^{n} a_m a_n$ for every n, and we deduce that (3.5), with n replaced by n_j, tends to zero with increasing j. This implies the first line of (3.4), with n_j replacing n, tends to zero while the second line of (3.4) is never positive for any n and any $f \in C$. These facts allow us to assert finally that

$$\text{(3.6)} \qquad \overline{\lim_j}(g_{n_j,r}-g) \cdot (g_{n_j,1}-f) \le 0, \quad f \in C.$$

We will need a slightly different version of (3.6), where $g_{n_j,1}$ is replaced by $g_{n_j,r}$, i.e.,

$$\text{(3.7)} \qquad \overline{\lim_j}(g_{n_j,r}-g) \cdot (g_{n_j,r}-f) \le 0, \quad f \in C.$$

This follows easily from (3.6) by writing $(g_{n_j,r}-g)\cdot(g_{n_j,1}-f)$ $= (g_{n_j,r}-g)\cdot(g_{n_j,1}-g_{n_j,r}) + (g_{n_j,r}-g)\cdot(g_{n_j,r}-f)$ and showing that the first term goes to zero as $j \longrightarrow +\infty$. This first term is bounded by $\|g_{n_j,r}-g\|\|g_{n_j,1}-g_{n_j,r}\|$ by the Cauchy-Schwarz inequality. Recall the equality (3.2), where all terms are nonnegative. Since $\|g_{nr}-g^*\|^2$ appears on the right side, the sequence $\{g_{nr}-g^*\}$, as well as the sequences $\{g_{nr}\}$ and $\{g_{nr}-g\}$, must be uniformly bounded. Hence, $\|g_{n_j,r}-g\|$ is uniformly bounded in j while from the remarks immediately following (3.3), we have $\|g_{n_j,1}-g_{n_j,r}\| \longrightarrow 0$ as $j \longrightarrow +\infty$, and hence (3.7) is established.

Now, by weak compactness and boundedness properties, the sequence $\{g_{n_j,r}\}$ must contain a subsequence converging weakly to $h \in H$. Clearly we may choose our original subsequence $\{n_j\}$ so that (3.7) holds as well as

$$g_{n_j,r} \xrightarrow{w} h, \quad j \longrightarrow +\infty. \tag{3.8}$$

Moreover, because $\|g_{n_j,r}\|$ is bounded we can actually choose a subsequence $\{n_j\}$ which satisfies the third condition

$$\|g_{n_j,r}\| \longrightarrow t \geq 0, \quad j \longrightarrow +\infty. \tag{3.9}$$

In summary, we have argued that a subsequence $\{n_j\}$ exists with (3.7)-(3.9) holding simultaneously.

We now use a result from Balakrishnan (1971), p. 19, which states that if $\{x_n\} \xrightarrow{w} x$ and $\|x_n\| \longrightarrow s$, then $\|x\| \leq s$. Hence, in the above, $\|h\| \leq t$ so that

$$\begin{aligned} 0 &\geq \overline{\lim_j}(g_{n_j,r}-g)\cdot(g_{n_j,r}-f) \\ &= t^2 - h\cdot f - h\cdot g + g\cdot f \geq \|h\|^2 - h\cdot f - h\cdot g + g\cdot f \\ &= (h-g)\cdot(h-f). \end{aligned}$$

We have therefore shown that $(g-h)\cdot(h-f) \geq 0$ for all $f \in C$. If we can verify that $h \in C$, then, by the optimality criterion (2.1), we will have $h = g^*$.

To this end we first claim that $g_{n_j,i} \xrightarrow{w} h$ for any $1 \leq i \leq r$. This is immediate once again from the remarks after (3.3) and from (3.8). Furthermore, applying another result from Balakrishnan (1971), p. 20, we have for each i a subsequence $\{j_k\}$ such that

$$\frac{1}{m}\left[g_{n_{j_1},i} + \cdots + g_{n_{j_m},i}\right] \xrightarrow{s} h, \quad m \longrightarrow +\infty.$$

However the sum on the left is a convex combination of elements in the convex set C_i for all m and is thus in C_i for all m. Since C_i is closed, we see that $h \in C_i$. But this is true for all i so that $h \in C$ and hence $h = g^*$.

Going back to (3.7) and setting $f = h$, we see that $\lim_j(g_{n_j,r}-g)\cdot(g_{n_j,r}-h) = t^2-\|h\|^2 \leq 0$. This coupled with an earlier result implies $t = \|h\|$. By Theorem 1.5 of Balakrishnan (1971), p. 20, we now deduce that $g_{n_j,r} \xrightarrow{s} h = g^*$.

We have not quite completed the proof, since we have exhibited only a subsequence of $\{g_{nr}\}$ converging strongly to g^*. However, in a manner similar to the derivation of (3.2) the following identity can be deduced.

$$\|g_{n_j,r}-g^*\|^2 = \|g_{n_j+k,r}-g^*\|^2 + \sum_{i=1}^{r}\sum_{s=1}^{k}\|I_{n_j+s-1,i}-I_{n_j+s,i}\|^2$$

(3.10)

$$+ 2\sum_{i=1}^{r}\sum_{s=1}^{k}(I_{n_j+s-1,i}-I_{n_j+s,i})\cdot(g_{n_j+s,i}-g^*).$$

This last double sum can be split into the difference of two double sums and the first reindexed to set (3.10) equal to

$$(3.11)\qquad 2\sum_{i=1}^{r}\sum_{s=0}^{k-1} I_{n_j+s,i}\cdot(g_{n_j+s+1,i}-g^*) \\ -\ 2\sum_{i=1}^{r}\sum_{s=1}^{k} I_{n_j+s,i}\cdot(g_{n_j+s,i}-g^*).$$

Now, both lines of (3.11) can be further transformed to

$$(3.12)\qquad 2\sum_{i=1}^{r} I_{n_j,i}\cdot(g_{n_j+1,i}-g^*)\ +$$

$$(3.13)\qquad 2\sum_{i=1}^{r} (-I_{n_j+k,i})\cdot(g_{n_j+k,i}-g^*)\ +$$

$$(3.14)\qquad 2\sum_{i=1}^{r}\sum_{s=1}^{k-1} I_{n_j+s,i}\cdot(g_{n_j+s+1,i}-g_{n_j+s,i}).$$

If we subtract and add $g_{n_j,i}$ inside the parentheses of (3.12), then (3.12) can be split into

$$(3.15)\qquad 2\sum_{i=1}^{r} I_{n_j,i}\cdot(g_{n_j+1,i}-g_{n_j,i})\ +$$

$$(3.16)\qquad 2\sum_{i=1}^{r} I_{n_j,i}\cdot(g_{n_j,i}-g^*).$$

Each term in (3.15) is nonnegative by (2.1). We can also easily verify that $(g_{n_j,r}-g)\cdot(g_{n_j,1}-g^*)$ tends to zero as $j \longrightarrow +\infty$, which says that (3.4), with n replaced by n_j and $f = g^*$, tends to zero. But the first line of (3.4) converges to zero, and we have the second line vanishing as $j \longrightarrow +\infty$, so that (3.16) tends to zero with increasing j. Additionally, each of the terms in (3.13) and (3.14) is nonnegative, again by (2.1). Therefore, all of the above implies that the term $\|g_{n_j,r}-g^*\|^2$ consists of nonnegative terms and a sum which tends to zero as $j \longrightarrow +\infty$. But this says that $\|g_{n_j+k,r}-g^*\|$ $\longrightarrow 0$ as $j \longrightarrow +\infty$ for any integer $k \geq 0$, since $\|g_{n_j,r}-g^*\|^2$ tends to zero, and the theorem is established.

Before concluding this section, we remark that if C_i is a subspace, then it is not necessary for the nth cycle to remove the increment $I_{n-1,i}$ before projecting onto C_i. This is simply because the projection operation, say P_i, onto C_i is linear and $P_i(I_{n-1,i}) = 0$. It should be clear that this fact also prevails when C_i is a translate of a subspace.

4. APPLICATIONS. We consider first the estimation of β in the linear model $Y = X\beta+\epsilon$, where Y is $n\times 1$, X is $n\times k$ of full rank and $\mathrm{Cov}(\epsilon) = \sigma^2\Omega$ with Ω assumed positive definite and known. Recall that the generalized least squares estimator of β is $\hat{\beta} = (X'\Omega^{-1}X)^{-1}X'\Omega^{-1}Y$, which is of course the unrestricted M.L.E if the errors are jointly normally distributed.

Frequently it is desired to place restrictions on the estimate of β. Specifically, restrictions of the form $\beta \in C_i = \{\beta \mid \sum_{j=1}^{k} a_{ij}\beta_j \underset{(=)}{\leq} b_i\}$ are often imposed. The inequality restriction places β in a half-space and the equality restriction forces β to lie in a $k-1$

dimensional hyperplane. These restrictions lead to the constrained maximum likelihood problem:

$$(4.1)\qquad \text{Minimize}\quad (Y-X\beta)'\Omega^{-1}(Y-X\beta)\quad \text{for}\quad \beta \in C_1\cap C_2\cap\cdots\cap C_r.$$

The usual manipulations on the objective function in (4.1) establish that it equals

$$(\beta-\hat{\beta})'X'\Omega^{-1}X(\beta-\hat{\beta}) + (Y'\Omega^{-1}Y-\hat{\beta}'X'\Omega^{-1}X\hat{\beta})$$

and hence problem (4.1) is equivalent to:

$$(4.2)\qquad \text{Minimize}\ (\beta-\hat{\beta})'X'\Omega^{-1}X(\beta-\hat{\beta}),\quad \text{for}\quad \beta \in C_1\cap C_2\cap\cdots\cap C_r.$$

a standard quadratic programming problem in $\mathbb{R}^k$.

The thing to note about problem (4.2) is that its solution is the projection of the unrestricted estimator $\hat{\beta}$ onto the intersection of closed convex sets in $\mathbb{R}^k$, where the inner product is defined as $\beta_1\cdot\beta_2 = \beta_1'X'\Omega^{-1}X\beta_2$. Hence the algorithm of section 2 may be useful in finding the restricted estimator $\tilde{\beta}$. It may be easier to make the transformation $\delta = P'\beta$ where P is an orthogonal matrix whose columns are eigenvectors of $X'\Omega^{-1}X$ with eigenvalues $w_1,\cdots,w_k > 0$, i.e., $P'X'\Omega^{-1}XP = \text{diag}(w_1,\cdots,w_k)$. Then problem (4.2) is transformed to:

$$(4.3)\qquad \text{Minimize}\ \sum_{j=1}^{k} w_j(\delta_j-\hat{\delta}_j)^2$$

$$\delta \in \bigcap_{i=1}^{r}\{\delta \mid \sum_{j=1}^{k} c_{ij}\delta_j \underset{(=)}{\le} b_i\}$$

with $\hat{\delta} = P'\hat{\beta}$ and $c_i' = a_i'P$. The solution to (4.2) is then $\tilde{\beta} = P\tilde{\delta}$ where $\tilde{\delta}$ solves (4.3), i.e., $\tilde{\delta}$ is a projection of $\hat{\delta}$ onto the intersection of closed convex sets.

When $r = 1$ in (4.3) we form the Lagrangian

$$\sum_{j=1}^{k} w_j(\delta_j-\hat{\delta}_j)^2 + \lambda\left[\sum_{j=1}^{k} c_j\delta_j-b\right],$$

where the subscript i has been dropped, and differentiate to obtain the equations:

$$2w_j(\delta_j-\hat{\delta}_j) + \lambda c_j = 0; \quad j = 1,2,\cdots,k$$

$$\sum_{j=1}^{k} c_j\delta_j = b.$$

These equations can be used to yield:

$$\tilde{\delta}_j = \begin{cases} \hat{\delta}_j & ; \quad \sum_{j=1}^{k} c_j\hat{\delta}_j \le b \\ \hat{\delta}_j-\lambda c_j/2w_j & ; \quad \sum_{j=1}^{k} c_j\hat{\delta}_j > b \end{cases}$$

where $\lambda = 2\left[\sum_{j=1}^{k} c_j\hat{\delta}_j-b\right]/\left[\sum_{j=1}^{k} c_j^2/w_j\right]$. Of course, if the constraint in (4.3) is an equality constraint always set $\tilde{\delta}_j = \hat{\delta}_j - \lambda c_j/2w_j$. Here we see that projections onto the component sets in (4.3) are easy to obtain, and therefore the algorithm of section 2 can be applied to (4.3) to obtain convergent approximations to $\tilde{\delta}$.

The point we wish to emphasize is that the algorithm furnishes a procedure which handles a fairly large class of restricted regression problems and which is simple to program. For specific examples of some isotonic regression problems and other related problems of interest, see Dykstra (1983). The algorithm has been applied to several of these problems and found to yield quite rapid convergence in most cases.

We will now discuss briefly a few applications of the algorithm we discovered in connection with the fitting of certain restricted minimization problems in the theory of splines. Much has been written in recent years regarding cubic smoothing splines, i.e., solutions to the problems:

$$(4.4) \qquad \text{Minimize} \sum_{i=1}^{n} [f(x_i)-y_i]^2 + \lambda\int_a^b (f''(x))^2 dx, \text{ for } f \in H_2[a,b]$$

where λ is fixed and positive and $H_2[a,b]$ is the set of functions on $[a,b]$ with absolutely continuous first derivative and square-integrable second derivative. For a good survey article including an extensive bibliography which discusses these and other classes of splines and their growing influence on statistics, see Wegman and Wright (1983). As mentioned in this article, the solution to (4.4) is often entertained as an attractive nonparametric estimator of the regression curve $E(Y|X=x) = f(x)$. For a more recent discussion of these matters, also see B.W. Silverman (1985). Note that no restriction has been placed on the solution to (4.4) other than it be in $H_2[a,b]$. Certainly one restriction which might be imposed on the solution is that it be convex. We have found the algorithm of section 2 to be quite effective in obtaining the solution to this restricted problem and we are in the process of writing up the details for publication. We remark that this restricted problem cannot be reduced to a finite-dimensional quadratic programming problem and, thus, the fitting of this convex smoothing spline constitutes an essential use of the algorithm in an infinite-dimensional setting.

In connection with the above we also note, once again withholding details, that the algorithm is well-suited to handle the closely related problem:

$$\text{(4.5)} \qquad \text{Minimize} \int_a^b (f''(x))^2 dx$$

$$f \in H_2[a,b] \cap W$$

where $W = \{f \mid f \text{ is convex}, f(x_i) = y_i; \ i = 1,\cdots,n\}$. The solution to (4.5) without the convexity constraint is, of course, the well-known cubic interpolating spline. See again Wegman and Wright (1983) for a discussion of this interpolating spline and some of its properties. It is considered appropriate if there is no noise in the data, i.e., an exact fit to the y_i's is desired. The solution to (4.5) with the convexity constraint is then the smoothest convex $H_2[a,b]$ function, in the sense of minimizing the integral in (4.5), which interpolates exactly to the data. Of course, it must be assumed that the set $H_2[a,b] \cap W$ in (4.5) is nonempty. The fitting of this convex interpolating spline is, like the fitting of the convex smoothing spline, not reducible to a finite-dimensional problem and is therefore another instance of the algorithm's usefulness in an infinite-dimensional context.

5. CONCLUDING REMARKS. In the previous section we have seen that the algorithm of Section 2 can be applied to the linear model to yield convergent approximations to restricted regression estimates. Also some applications to spline smoothing have been mentioned which take place in an infinite-dimensional setting. We feel that these applications of the proposed algorithm are but a few among potentially many since the algorithm has been shown to work in any Hilbert space with the only restriction on the projection regions being that of convexity. Consequently we continue to look for new and interesting applications in both finite and infinite-dimensional spaces.

Regarding the convergence of the algorithm, note that (3.2) consists only of nonnegative terms and that g_{ni} converges to g^* in a monotone manner (in that $\|g_{ni}-g^*\|$ is nonincreasing as n and i increase). We recommend the use of $\frac{1}{r}\sum_{i=1}^{r} g_{ni}$ as an estimate of g^* rather than just g_{nr} since this damps out some of the oscillations as the projections move from boundary to boundary.

Finally, an open question is whether a natural modification of the algorithm would converge correctly if one had a countably infinite intersection of convex constraints. In particular, one might employ the algorithm for some fixed, finite number of constraints on the first cycle, and then add an additional number of constraints on each subsequent cycle of the algorithm. Every constraint would eventually be considered, and one would hope that the monotonicity properties of the algorithm would lead to the correct convergence.

REFERENCES

Balakrishnan, A.V. (1971). *Introduction to Optimization Theory in a Hilbert Space*. Springer-Verlag, Berlin.

Barlow, R.E., Bartholomew, D.J., Bremner, J. M. and Brunk, H.D. (1972). *Statistical Inference Under Order Restrictions*. John Wiley and Sons, New York.

Barlow, R.E. and Brunk, H.D. (1972). "The isotonic regression problem and its dual." *J. Amer. Statist. Assoc.* 67: 140-147.

Dykstra, R.L. (1983). "An algorithm for restricted least squares regression." *J. Amer. Statist. Assoc.* **78**: 837-842.

Von Neumann, J. (1950). *Functional Operators* (Vol. II). Princeton University Press, Princeton, N.J.

Wegman, E.J., Wright, I.W. (1983). "Splines in statistics." *J. Amer. Statist. Assoc.* **78**, 351-365.

Wiener, N. (1955). "On factorization of matrices." *Comm. Math. Helv.* **29**: 97-111.

James P. Boyle
Department of Statistics and Actuarial Science
University of Iowa
Iowa City, IA 52242

Richard L. Dykstra
Department of Statistics and Actuarial Science
University of Iowa
Iowa City, IA 52242

ISOTONIC M-ESTIMATION

Sue Leurgans
The Ohio State University

AMS 1980 subject classifications: 62G99, 62F35.
Key words and phrases: recursive partitioning, robustness, isotonic regression, order restricted inference.

ABSTRACT

Robust partitioning algorithms for isotonic regression are shown to have anomalous behavior.

1. **INTRODUCTION.** Most statistical inferences are based on probability models; often on parametric models. Since the assumptions of any probabilistic model are usually abstract postulates rather than empirically determined facts, the sensitivity of the inference under small changes needs to be understood. Robust methods which are insensitive to the exact details of parametric models have been the object of much research, and are beginning to be applied.

In this paper, one of the standard robust methods is reviewed for use in order-restricted inference. Magel and Wright (1984) discuss robust pooling algorithms when many replicates are available. When design points are not replicated, pooling algorithms have the disadvantage of not being robust until several points have been pooled. Therefore partitioning algorithms would seem desirable in such contexts. However, those recursive partitioning algorithms which resemble those of Gordon and Olshen (1978, 1980) are shown by example to fail for some robust estimators. The implications for model selection are indicated briefly.

In the remainder of this section, some of the notation is introduced. In the next section, the M-estimation approach to robust estimation is reviewed and further notation introduced. In the third section, several of the available algorithms are described. The fourth section gives a specific numerical example and the final section summarizes the implications of this example. The example of the fourth section is neither a pathology nor a concidence—the fifth section contains a general construction of the example of section 4.

This paper assumes that n independent pairs of observations (x_i, Y_i) are available, where the distribution of Y_i depends on a location parameter θ_i, which is assumed to be an isotone function of x_i : $\theta_i = \theta(x_i)$. The x's will be referred to as the design points, the Y's will be referred to as the values, and the pairs will be referred to as points. Nuisance parameters may be present, but are not assumed to have any structure. For ease of exposition, the x_i are taken to be totally ordered and equally weighted, so that it suffices to assume that $x_1 \le x_2 \le \cdots < x_n$ and that $\theta_1 \le \theta_2 \le \cdots \le \theta_n$.

2. M-ESTIMATION. The method of M-estimation is widely applicable. A criterion, which can be thought of as measuring the adequacy of a particular value of the parameter, is defined. The M-estimator is the value of the parameter at which the criterion function is minimized.

The standard form of the M-estimation criterion, thought of as a function of the n-fold vector of possible location parameters is

$$Q(\underset{\sim}{\theta}) = \sum_{i=1}^{n} \rho(Y_i - \theta_i).$$

If ρ is symmetric about 0 with $\rho(0) = 0$, and if $\rho(x)$ increases as x moves away from 0, then the ith term in the summation can be thought of as a measure of the size of the ith residual. The isotonic M-estimator is the vector of ordered values $t_1 \leq \cdots \leq t_n$ minimizing $Q(\underset{\sim}{\theta})$ over all nondecreasing θ_i's. The estimator can be extended to a function θ on the design space in any manner preserving the order assumptions. If $\rho(y) = y^2$, the isotonic M-estimator is the isotonic regression. If $\rho(y) = |y|$, the isotonic M-estimator is the isotonic median. If the distribution of the Y_i's are translates of a location family with density f and if $\rho(y) = -\log(f(y))$, the isotonic M-estimator is the restricted maximum likelihood estimator of the location parameters. However, the function ρ can be chosen without reference to the underlying density.

The counterexamples below use the one-parameter family of functions

$$\rho(x) = \ln(1+e^{cx}) - cx/2 - \ln(2)$$

with derivative

$$\psi(x) = c\left\{\frac{e^{cx}}{1+e^{cx}} - \frac{1}{2}\right\}.$$

Note that $(\psi(x)/c + 1/2)$ is the logistic distribution function. The scaling constant c here is the reciprocal of the usual tuning constant. For small c, $\psi(x)$ will be nearly linear for a relatively wide interval around the origin, and M-estimation will be close to least-squares estimation. For large c, $\psi(x)$ will nearly attain its limits $\pm c/2$ outside of a relatively narrow interval around the origin, and M-estimation will be close to minimum absolute deviation estimation. These ψ functions can also be thought of as smooth approximations to the piecewise linear ψ function popularized by Huber (Holland and Welsch (1977)).

Note that M-estimators are not equivariant under linear transformations of the Y's unless an estimated scale parameter is introduced in the denominator of the argument of ρ. As this feature, while important in practice, does not affect the issues raised below, it will be neglected here.

Because active constraints force a monotone function to have the same value for at least two design points, the isotonic M-estimator is constant on sets (possibly singleton sets) of design points. These sets will be called the solution sets; the symbol S will be reserved for solution sets. If the solution sets are known to be $S_1, S_2, \cdots, S_J$, then the criterion function can be written as the sum of J criterion functions: one function for each solution set. More specifically, if

$$\rho(t;A) = \sum_{x_i \in A} \rho(Y_i - t),$$

then

$$Q(t_1, \cdots, t_n) = \sum_{j=1}^{J} \rho(t_j; S_j).$$

Thus, if the solution sets were known, the computation of the isotonic M-estimator would require the computation of J univariate M-estimators. Computation of the isotonic regression therefore requires determination of the solution sets.

If ρ is differentiable with derivative ψ, the minimizer of $\rho(t;B)$ must be a value t which satisfies

$$\psi(t;B) = \sum_{x_i \in B} \psi(Y_i - t) = 0.$$

If ρ is strictly convex, this characterizing equation will have a unique solution, denoted by $\psi(B)$. For discussions of computational strategies, see Hogg (1979), Bickel (1976), and Huber (1981).

Visualizing the graph of $\psi(t;B)$ provides an easy view of the properties of $\psi(B)$. If ρ is strictly convex, $\psi(Y_i - t)$ is decreasing, and so is $\psi(t;B)$ for all nonempty sets B. The value of t at which $\psi(t;B)$ crosses 0 is $\psi(B)$. If B_1 and B_2 are two disjoint sets of design points, $\rho(t;B_1 \cup B_2) = \rho(t;B_1) + \rho(t;B_2)$. Since the point at which the sum of two decreasing functions crosses 0 cannot lie outside the crossing points of the individual functions, $\psi(B_1 \cup B_2)$ cannot lie outside the interval with endpoints $\psi(B_1)$ and $\psi(B_2)$ as long as ρ is convex. This pooling property, known as the Cauchy Mean Value Property, will be used below. Note that if ψ is not monotone, the Cauchy Mean Value Property will not hold for the corresponding isotonic M-estimators, including the popular re-descenders whose ψ functions eventually approach zero as their arguments diverge.

Two sample properties of M-estimation should be noted. If n is one, the M-estimator is Y_1. If n is 2,

$$\rho(t;B) = \rho(Y_1 - t) + \rho(Y_2 - t).$$

If ρ is symmetric about 0, $\rho(t,B)$ is symmetric about $(Y_1+Y_2)/2 = \psi(B)$. The function ρ need not be known: the M-estimator will be the same as the least-squares estimator unless three or more data points are used.

3. CLASSIFICATION OF ALGORITHMS. For M-estimators with increasing ψ function, the closed form formula of Robertson and Wright (1974) applies. The formula uses upper and lower sets. In a totally ordered space, a lower set L is any set of consecutive design points which includes the smallest design point, and an upper set U is any set of consecutive design points which includes the largest design point. The symbol L will be reserved for lower sets and the symbol U for upper sets below. The formula for the isotonic M-estimator at a design point x is

$$\theta^*(x) = \max_{x \in L} \min_{x \in U} \psi(L \cap U).$$

While this formula *is* explicit, many M-estimates have to be computed.

Other algorithms compute the isotonic M-estimators from solution sets. These algorithms can be classified as either amalgamation or partitioning algorithms.

Amalgamation algorithms gradually build up the solution sets until the estimates from the pooled design points are monotone. If computation is halted before the procedure is complete, the estimates based on the current sets will not be monotone. These algorithms resemble backwards elimination methods of model selection for multiple regression in which superfluous predictors are gradually eliminated. For totally ordered designs, the Up-and-Down Blocks (Barlow, Bartholomew, Bremner and Brunk, p. 72) and the Pool-Adjacent Violators (Barlow et al., p. 13) algorithms apply.

Partitioning algorithms gradually split the design until solution sets are obtained. This process is akin to forward selection for multiple regression, in which a simple model is gradually augmented by inclusion of effective predictors. Recursive partitioning algorithms are also attractive because of their recent application to classification and nonparametric regression (Breiman, Friedman, Olshen and Stone (1984)). Recursive partitioning rules divide the design space into disjoint blocks. The design space is the initial block. The algorithm consists of a rule which determines whether a given block should be split and a procedure which splits each block into two subblocks. The rule is then applied to the resulting blocks until no blocks will be split. A well-designed algorithm will always terminate after a finite number of steps. When the partitioning has been completed, the regression function on a block is estimated by the value of some location estimator applied to the values in the block.

Several recursive partitioning algorithms terminate in the isotonic solution. The minimal upper set and maximal lower set algorithms shave off the solution sets one at a time from either the lowest or the highest edge. The block B is split into L′ and U′, where L′ is the lower set containing the most design points among those lower sets L satisfying

$$\psi(L) = \min_{L \subset B} \psi(L).$$

One advantage of partitioning algorithms is that if these algorithms are terminated early, the M-estimates of the current sets are monotone.

Operationally, one disadvantage of the minimal lower and maximal upper set algorithms is that stopping these algorithms before completion will yield a truncated version of the isotonic M-estimator, in the sense that the intermediate estimates equal the isotonic M-estimator at some of the design points and are constant on the rest of the design points. A coarse version of the isotonic estimator which is close to the M-estimator at many design points would be preferable. Furthermore, the number of decompositions examined will be smaller if the early subblocks could be of comparable magnitude. Therefore it is desirable to obtain other partitioning algorithms for isotonic M-estimation.

One natural candidate is the direct analog of the forward selection algorithm for multiple regression. Such an algorithm would split the block B into the largest lower set L and its complement in B (U) such that

$$\rho(\psi(L);L) + \rho(\psi(U);U) = \min[\rho(\psi(L');L') + \rho(\psi(U');U')].$$

Since the criterion function for the estimator corresponding to this partition will lower the criterion as much as possible by splitting the block under consideration, this algorithm will be referred to as the criterion reduction algorithm. The next section demonstrates that this algorithm can fail.

4. COUNTEREXAMPLE FOR PARTITIONING ALGORITHMS. In this section, a counterexample to criterion reduction algorithms for M-estimation is presented. A reflection function is defined which is nonlinear if the distance from one value to the values of a set of data points is an asymmetric function of the single value. The reflection functions for several examples illustrate the possibilities. In the next section, the behavior of the reflection function is shown to determine whether the criterion-reduction algorithm can split solution sets.

Because M-estimation is least-squares estimation unless more than two points are used, the simplest counterexamples will use sets S_1 and S_2 of size 3 with asymmetric values. Take S_1 to have three points with $x_1 < x_2 < x_3$ and $Y_1 = 4$, $Y_2 = 3$, $Y_3 = 0$. Take S_2 to have three points with $x_4 < x_5 < x_6$ and $Y_4 = 5$, $Y_5 = 2$, and $Y_6 = 1$. If $x_4 > x_3$, S_1 and any lower points form a lower set, and S_2 and any upper points form an upper set. It will be demonstrated below that for some ρ, although S_1 and S_2 are subsets of a solution set, they will be separated by a criterion reduction algorithm.

To demonstrate the problem explicitly, add four points below x_1 with value 0 and four points above x_6 with values 5. Thus the 14 data pairs can be taken to be the set {(1,0),(2,0),(3,0),(4,0),(5,4),(6,3),(7,0),(8,5),(9,2),(10,1),(11,5),(12,5),(13,5),(14,5)}. The solution sets are fairly easy to determine using one step of the minimal lower set or maximal upper set algorithm. The first four points and the last four points each constitute solution sets. If a Pool Adjacent Violator algorithm is applied to the inner six points, the points in S_1 and in S_2 will be pooled into those two sets. If $\psi(S_1) < \psi(S_2)$, S_1 and S_2 are solution sets, otherwise $S_1 \cup S_2$ is a solution set. Thus the recursive partitioning algorithm fails if Q is minimized by a split which separates S_1 and S_2 and if c is such that $\psi(S_1) \geq \psi(S_2)$.

All possible splits (L,U) are tabulated in Table 4.1 by n(L), the number of points in L for c = .1, 1, 10, ∞. The sets S_1 and S_2 will be separated if Q is minimized when n(L) = 7. If Q is smallest when n(L) is 4 or 7 , S_1 and S_2 will not be separated. This data set is preserved if the design points are reflected around 7.5 and the values are reflected about 2.5. Therefore, for each c, the $\rho(u;U)$ entries are the $\rho(\ell;L)$ entries in reverse order. Also, the sums of the first $\psi(L)$ entry and the last $\psi(U)$ entry, of the

Table 4.1. Summary of all upper/lower decompositions for example discussed in Section 4, tuning constant $c = .1, 1.0, 10, \infty$. See text for notation.

$c = .1$

$n(L)$	$\psi(L)$	$\psi(U)$	$\rho(\ell;L)$	$\rho(u;U)$	Q
1	0	2.69	0	.0758	.0758
2	0	2.92	0	.0660	.0660
3	0	3.19	0	.0544	.0544
4	0	3.50	0	.0405	.0405
5	.79	3.45	.0159	.0402	.0561
6	1.16	3.50	.0210	.0399	.0609
7	1.00	4.00	.0224	.0224	.0448
8	1.50	3.84	.0399	.0210	.0609
9	1.55	4.21	.0402	.0159	.0561
10	1.50	5.00	.0405	0	.0405
11	1.81	5.00	.0405	0	.0544
12	2.08	5.00	.0660	0	.0660
13	2.31	5.00	.0758	0	.0758

$c = 1.0$

$n(L)$	$\psi(L)$	$\psi(U)$	$\rho(\ell;L)$	$\rho(u;U)$	Q
1	0	2.80	0	6.24	6.24
2	0	3.12	0	5.40	5.40
3	0	3.44	0	4.42	4.42
4	0	3.75	0	3.29	3.29
5	.48	3.70	1.21	3.28	4.49
6	.90	3.86	1.77	3.20	4.97
7	.73	4.27	1.85	1.85	3.69
8	1.14	4.10	3.20	1.77	4.97
9	1.30	4.52	3.28	1.21	4.49
10	1.25	5.00	3.29	0	3.29
11	1.56	5.00	4.42	0	4.42
12	1.88	5.00	5.40	0	5.40
13	2.20	5.00	6.24	0	6.24

$c = 10$

$n(L)$	$\psi(L)$	$\psi(U)$	$\rho(\ell;L)$	$\rho(u;U)$	Q
1	0	3.00	0	121.68	121.68
2	0	3.50	0	106.70	106.70
3	0	4.00	0	88.07	88.07
4	0	4.42	0	68.10	68.10
5	.05	4.78	19.18	65.39	84.57
6	.11	4.86	33.09	56.96	90.05
7	.08	4.92	33.20	33.20	66.41
8	.14	4.89	56.96	33.09	90.05
9	.22	4.95	65.39	19.18	84.57
10	.58	0	68.10	0	68.10
11	1.00	0	88.07	0	88.07
12	1.50	0	106.70	0	106.70
13	2.00	0	121.68	0	121.68

$c = \infty$

$n(L)$	$\psi(L)$	$\psi(U)$	$\rho(\ell;L)$	$\rho(u;U)$	Q
1	0	3	0	26	26
2	0	3.5	0	23	23
3	0	4	0	19	19
4	0	4.5	0	15	15
5	0	5	4	14	18
6	0	5	7	12	19
7	0	5	7	7	14
8	0	5	12	7	19
9	0	5	14	4	18
10	.5	5	15	0	15
11	1	5	19	0	19
12	1.5	5	23	0	23
13	2	5	26	0	26

second $\psi(L)$ entry and the second-to-last $\psi(U)$ entry, and so on, are 5.

When $c = .1$, $\psi(S_1) = 2.34 < 2.66 = \psi(S_2)$, and the sets S_1 and S_2 are solution sets. When $c = 1$, $\psi(S_1) = 2.53 > 2.47 = \psi(S_2)$, and thus S_1 and S_2 are not solution sets. The lowest value of Q is 3.29, which occurs when $n(L)$ is 4 or 10, so S_1 and S_2 are not separated. The criterion-reduction algorithm works here, although the next section shows that S_1 and S_2 can be separated if more points are added. When $c = 10$, $\psi(S_1) = 3 > 2 = \psi(S_2)$, so that S_1 and S_2 are not solution sets. However, because the lowest value of Q (66.41) occurs when $n(L)$ is 7, the criterion reduction algorithm separates S_1 and S_2 and splits the solution set S. When $c = \infty$ and the sum of absolute deviations is used to define Q, $\psi(S_1) = 3 > 2 = \psi(S_2)$, and the criterion reduction algorithm fails again.

The next section contains a proof that these counterexamples are possible because $\rho(x;B)$ can be an asymmetric function of x unless ρ is quadratic. For fixed B, $\rho(x;B)$ can be thought of as

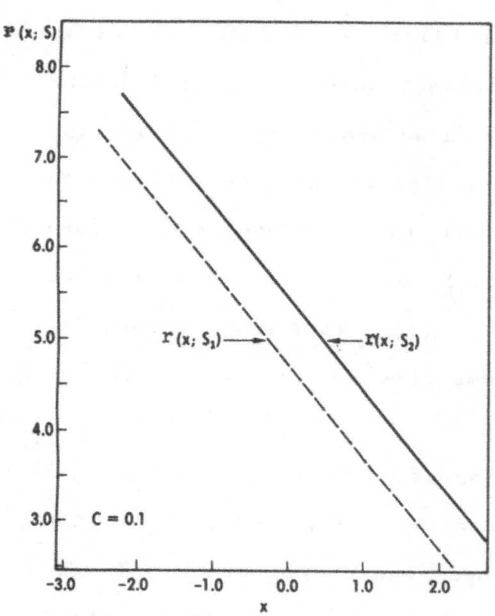

Figure 4.1. Approximations to $r(x;S_j)$, $j = 1,2$, $c = .1$.

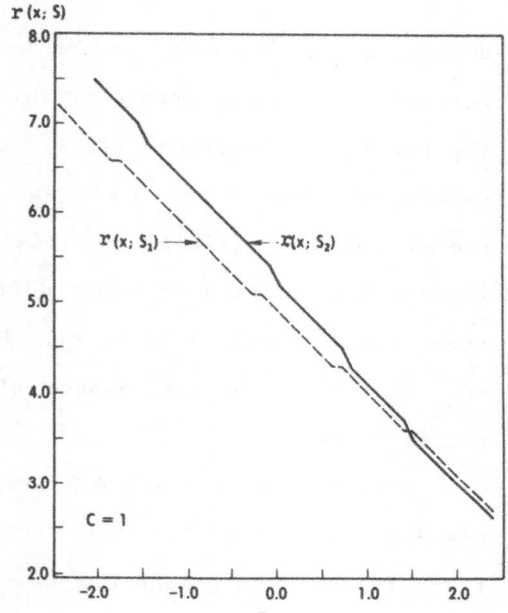

Figure 4.2. Approximations to $r(x;S_j)$, $j = 1,2$, $c = 1$.

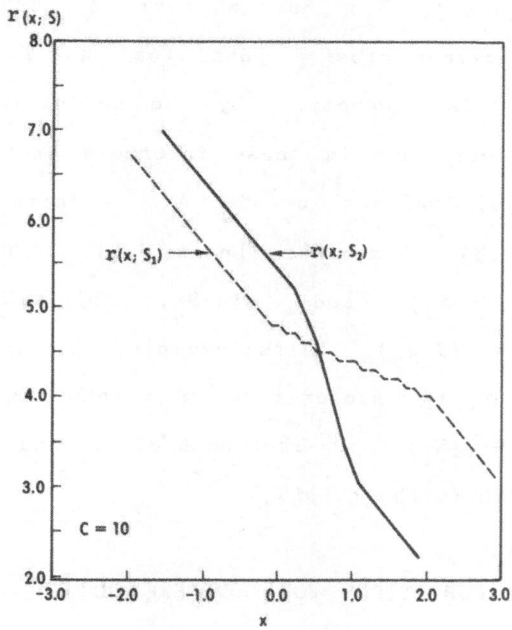

Figure 4.3. Approximations to $r(x;S_j)$, $j = 1,2$, $c = 10$.

measuring the distance between a single value x and the values in the set B. To see asymmetry in the distance measures $\rho(x;A)$, define the reflection function $r(x;A)$ as the value above $\psi(A)$ having the same ρ-distance from $\psi(A)$ as x. For convex ρ, it suffices to require that $\rho(r(x;A);A) = \rho(x;A)$ and that $r(x;A) \geq x$. (See Theorem 5.1 for an equivalent definition.) If $\rho(x;A)$ is symmetric, then $r(x;A) = \psi(A) + (\psi(A)-x))$ for $x \leq \psi(A)$. Thus any nonlinearity of $r(x;A)$ is a consequence of asymmetries in the values of the points in A.

Figures 4.1, 4.2 and 4.3 display approximations to $r(x;S_1)$ and $r(x;S_2)$ for $c = .1$, $c = 1$ and $c = 10$, respectively. In the first figure, the graphs are nearly linear and do not cross. In the second figure, the functions are less linear and cross. In the third figure, the reflection functions cross near .6. Because $\psi(S_2) = 2 < 3 = \psi(S_1)$, $r(x;S_2) < r(x;S_1)$ for x near 2. Near $x = 1.6$, the two reflection functions cross. Thus, for $\ell < 1.5$, the interval $(r(\ell;S_1), r(\ell;S_2))$ is nonempty. By the definition of reflection functions, all values u in these intervals exceed $\psi(S_1)$. For $\ell = 0$, the interval includes $u = 5$. As x increases or decreases from $\psi(B)$, $\rho(x;B)$ increases (Lemma 5.2). Therefore $\rho(\ell;S_1) = \rho(r(\ell;S_1);S_1) < \rho(u,S_1)$ and $\rho(u;S_2) < \rho(r(\ell;S_2);S_2) = \rho(\ell;S_2)$ when $r(\ell;S_1) < u < r(\ell;S_2)$. In the example, 5 is ρ-closer to S_2 than 0 is and 0 is ρ-closer to S_1 than 5 is, even though $0 < 2 = \psi(S_2) < 3 = \psi(S_1) < 5$. Thus each of 0 and 5 is ρ-closest to the set S_j with furthest $\psi(S_j)$.

5. NECESSARY CONDITIONS FOR COUNTEREXAMPLES. In this section, the behavior of the criterion-reduction algorithm is studied mathematically when ρ is convex. Theorem 5.1 gives a sufficient condition for the criterion reduction algorithm not to split a solution set in subsets S_1 and S_2. If the sufficient condition can be established for all pairs of subsets of solution sets, then no solution sets can be split by the criterion reduction algorithm. The conclusion that the criterion reduction algorithm terminates with the isotonic M-estimator would follow, because it is easy to check that no solution sets will be left together. Lemma 5.3 establishes the appropriate properties for the sum-of-squares quadratic function, and Corollary 5.4 is that the criterion-reduction algorithm is an algorithm for isotonic regression.

The sufficient condition of Theorem 5.1 is violated if the reflection functions of S_1 and S_2 cross. If the functions cross, Theorem 5.5 shows that the rest of the data can be chosen so that the solution set containing S_1 and S_2 will be split. The numerical counterexamples of the preceding section show that the premise of Theorem 5.5 is not vacuous.

THEOREM 5.1. *If* (L,U) *splits a solution set* S *into* $S_1 = L \cap S$ *and* $S_2 = U \cap S$, *the criterion-reduction algorithm will not select* (L,U) *if*

$$u \le r(\ell;S_1) \quad \text{or} \quad r(\ell;S_2) \le u, \tag{5.1}$$

where $\ell = \psi(L)$, $u = \psi(U)$, *and*

$$r(x;A) = \text{supremum}\{y : \rho(y;A) \le \rho(x;A)\}.$$

Some properties of $r(x;A)$ will be useful in the proof of the theorem and the interpretation of the conditions. The properties, stated as Lemma 5.2, are immediate consequences of the convexity of ρ.

LEMMA 5.2. *The function* $r(x;A)$ *is decreasing on* $(-\infty,\psi(A))$ *and is the identity function on* $[\psi(A),\infty)$. *If* $\ell \le x \le r(\ell;A)$, *then* $\rho(\ell;A) \ge \rho(\ell;A)$. *If* $r(\ell;A) \le x$, *then* $\rho(\ell;A) \le \rho(x;A)$.

PROOF OF THEOREM 5.1. The theorem follows if the conditions imply that splits with S entirely in the upper set or entirely in the lower set have a lower criterion than the split (L,U). Of course, the split (L,U) is not a candidate split unless $\ell \le u$.

The split with S entirely in the upper set is preferable if

$$\rho(\ell;L) + \rho(u;U) \ge \rho(\ell_1;L_1) + \rho(\psi_2;S \cup U_2),$$

where $L_1 = L \cap S^c$, $U_2 = U \cap S^c$, $\ell_1 = \psi(L_1)$ and $\psi_2 = \psi(S \cup U_2)$. This inequality is equivalent to the positivity of

$$(5.2) \qquad [\rho(\ell;L_1)-\rho(\ell_1;L_1)]+[\rho(\ell;S_1)-\rho(u;S_1)] +[\rho(u;S \cup U_2)-\rho(\psi_2;S \cup U_2)].$$

The first and last terms in square brackets ([ ]) are positive by the definitions of ℓ_1 and of ψ_2, respectivley, when ρ is strictly convex. (If ρ is convex, but not strictly convex, then the terms are nonnegative.) Since $\ell \le u$, the first inequality of (5.1) and Lemma 5.2 imply the nonnegativity of (5.2).

Similarly, the split with S entirely in the lower set is preferable if

$$(5.3) \qquad [\rho(u;U_2)-\rho(u_2;U_2)] + [\rho(u;S_2)-\rho(\ell;S_2)] + [\rho(\ell;L_1 \cup S)-\rho(\psi_1;L_1 \cup S)]$$

is positive, where $u_2 = \psi(U_2)$ and $\psi_1 = \psi(L_1 \cup S)$. The first and third terms are again positive. The inequalities $\ell \le u$ and $r(\ell;S_2) \le u$ and Lemma 5.2 thus give the desired inequality.

If the criterion function is the sum of squares function, $r(x;A)$ can be derived explicitly.

LEMMA 5.3. *If* $\rho(x) = x^2$, *then*

$$r(x;A) = \begin{cases} \psi(A) + (\psi(A)-x) & x \le \psi(A) \\ x & x \ge \psi(A) \end{cases}.$$

PROOF. Because convex polynomials of fixed degree are closed under translation and addition, $\rho(x;A)$ is a convex quadratic function. Therefore $\rho(x;A)$ is symmetric about its unique minimum at $\psi(A)$. For x less than $\psi(A)$, $r(x;A)$ is the value y greater than $\psi(A)$ such that $\rho(x;A) = \rho(y;A)$. By the symmetry of $\rho(\cdot;A)$, this y must be the same distance above $\psi(A)$ as x is below $\psi(A)$. The lemma follows.

COROLLARY 5.4. *The criterion reduction algorithm applied to the sum-of-squares criterion for a total order terminates in the isotonic regression.*

PROOF. By Lemma 5.3, the graph of $r(x;A)$ is a continuous V-shaped function which coincides with the identity function above $\psi(A)$ and has slope -1 below $\psi(A)$. Because S is a solution set, but S_1 and S_2 are not, $\psi(S_1) \geq \psi(S_2)$. The implicit sketch verifies that $r(\ell;S_1) \geq r(\ell;S_2)$ for all ℓ and therefore all values u must satisfy at least one of the inequalities (5.1). Thus no solution sets can ever be split by the criterion reduction algorithm. Moreover, if $B = L \cup U$, where L and U are unions of solution sets,

$$\rho(\psi(B);B) = \rho(\ell;L)+\rho(u;U)+(\ell-\psi(L))^2 n(L)+(u-\psi(U))^2 n(U) > \rho(\ell;L)+\rho(u;U),$$

and any block containing more than one solution set will be split. Therefore the algorithm stops only when each block is a solution set.

If S is a solution set, the minimal lower sets algorithm implies that $\psi(S_1) > \psi(S_2)$. Lemma 5.2 then implies that $r(\ell;S_2) = \ell$ for ℓ in $[\psi(S_2),\psi(S_1))$ and $r(\ell;S_1) > r(\ell;S_2)$ on this interval, and thus that $r(\ell;S_1) > r(\ell;S_2)$ on this interval. If the two reflection functions do not cross, $r(\ell;S_1) > r(\ell;S_2)$ for all ℓ and at least one of the inequalities (5.1) holds for any u. Theorem 5.2 would then imply S_1 and S_2 will not be separated by the criterion-reduction algorithm. However, if the reflection functions cross, there will be some $\ell < u$ for which (5.1) does not hold. Theorem 5.5 shows that for such (ℓ,u), data sets exist for which the solution set S will be split.

THEOREM 5.5. *If both inequalities* (5.1) *are violated for some* $\ell < u$, *then* L_1 *and* U_2 *can be chosen so that* S *will be split.*

The proof uses L_1 and U_2 consisting of n_1 and n_2 design points at which the same values $\ell_1(n_1)$ and $u_2(n_2)$ are recorded. If n_1 and n_2 are large enough, Lemma 5.6 establishes that S_1 and

S_2 are dominated by L_1 and U_2.

LEMMA 5.6. *Let* B *be a fixed set of data points and let* a *be a fixed number. For each* n, A_n *is a set of* n *data points, all with the same value* a(n), *satisfying* $\psi(A_n \cup B) = a$. *Then*

$$\rho(a;A_n) = n\rho(a(n)-a) = c/n + o(n^{-1}),$$

where $c = \psi^2(a;B)/(2\psi'(0))$.

PROOF. The condition $\psi(A_n \cup B) = a$ implies that $\psi(a(n)-a) = -\psi(a;B)/n$. Because ψ has two bounded derivatives at 0, $\psi(a;B)/n = (a(n)-a)\psi'(0) + O((a(n)-a)^2)$ or $a(n)-a = \psi(a;B)/(n\psi'(0)) + O((n^{-2})$. Substitution of the above expression in a two-term Taylor series expansion for $\rho(a;A_n) = n\rho(a(n)-a)$ gives the conclusion desired.

PROOF OF THEOREM. It suffices to show that both (5.2) and (5.3) are negative. Because $S \cup U_2$ is equal to the disjoint union of the sets S_1, S_2 and U_2, $\rho(x;S \cup U_2) = \rho(x;S_1) + \rho(x;S_2) + \rho(x;U_2)$. Substituting this equation with $x = \psi_2$ and $x = u$ in (5.2) gives (5.4):

$$(5.4) \quad [\rho(\ell;S_1)-\rho(\psi_2;S_1)] + [\rho(u;S_2)-\rho(\psi_2;S_2)] + [\rho(\ell;L_1)-\rho(\ell_1;L_1)] + [\rho(u;U_2)-\rho(\psi_2;U_2)]$$

and (5.3) can be written as

$$(5.5) \quad [\rho(u;S_2)-\rho(\psi_1;S_2)] + [\rho(\ell;S_1)-\rho(\psi_1;S_1)] + [\rho(u;U_2)-\rho(u_2;U_2)] + [\rho(\ell;L_1)-\rho(\psi_1;L_1)].$$

By Lemma 5.2, the hypotheses of the theorem imply that

$$\epsilon = -\max\{[\rho(\ell;S_1)-\rho(u;S_1)],[\rho(u;S_2)-\rho(\ell;S_2)]\}$$

is positive. The continuity of $\rho(\ell;S_1)-\rho(\cdot;S_1)$ guarantees that there exists a positive constant δ_1 such that $|\psi_2-u| < \delta_1$ implies $\rho(\ell;S_1)-\rho(u;S_1) < -4\epsilon/5$. Similarly, there is a positive δ_2 such that whenever $|\psi_1-\ell| < \delta_2$, $\rho(u;S_2)-\rho(\psi_1;S_2) < -4\epsilon/5$. Therefore the first bracketed terms in (5.4) and in (5.5) are less than $-4\epsilon/5$. By the continuity of $\rho(\cdot;S_2)$ and $\rho(\cdot;S_1)$, there exist positive numbers δ_3 and δ_4 such that $|u-\psi_2| < \delta_3$ and $|\ell-\psi_1| < \delta_4$ imply that the second terms in (5.4) and in (5.5) are less than $\epsilon/5$. With $B = S_1$, $a = \ell$, and $A_n = L_1$, Lemma 5.6 implies that there exists N_1 such that $n_1 > N_1$ ensures that $\rho(\ell;L_1) < \epsilon/5$. Since $\rho(\ell;L_1) = 0$ by construction and $\rho(\psi_1;L_1) \geq 0$, for such n_1, the sum of the last two terms in (5.4) is less than $2\epsilon/5$. Since $\ell_1(n)$ and ψ_1 converge to ℓ, N_1 can be chosen large enough so that $|\psi_1-\ell| < \min(\delta_2,\delta_4)$, which implies that (5.4) $< -4\epsilon/5 + 3\epsilon/5 < 0$. Similarly, N_2 can be chosen large enough so that $\rho(u;U_2) < \epsilon/5$ and $|\psi_2-u| < \min(\delta_1,\delta_3)$, implying (5.5) $< -\epsilon/5 < 0$. The theorem follows.

6. **DISCUSSION**. The fact that the criterion-reduction algorithm is not applicable seems to present a challenge to robust inference in the presence of order restrictions. If replicates are unavailable, pooling algorithms will not be robust. If partitioning algorithms are

used, it appears that robustness can change how an individual point enters the analysis—causing a value to be more or less down-weighted according to how much it appears to be an outlier. However, down-weighting points that have high or low values will change the evidence they supply about the choice of solution sets.

It is not clear that these difficulties do not extend to model selection in general. Huber (1981, section 7.10) recommends fitting a large model and then doing inference within that model. This corresponds to backward elimination and amalgamation methods. Thus, while partitioning rules are consistent, partitioning rules cannot be expected to lead to the optimal solution, especially if one-sided constraints, such as order restrictions or positivity constraints are present. And such constraints are often present when the plausibility of a model is judged by attempting to interpret the coefficients. This suggests that robust model selection may be even more problematic than in familiar regression, in that the robust methods need to entertain a model in order to down-weight points. As various models are tried, the apparent needs for down-weighting can change drastically enough to cause confusion.

REFERENCES

Bickel, P.J. (1976). Another look at robustness: A review of reviews and some new developments. *Scand. J. Statist.* 3, 145-168.

Breiman, L., Friedman, J.H., Olshen, R.A. & Stone, C.J. (1984). *Classification and Regression Trees.* Wadsworth: Belmont, California.

Gordon, L. & Olshen, R.A. (1978). Asymptotically efficient solutions to the classification problem. *Ann. Statist.* 6, 515-533.

Gordon, L. & Olshen, R.A. (1980). Consistent nonparametric regression from recursive partitioning schemes. *J. Mult. Anal.* 10, 611-627.

Hogg, R.V. (1979). Statistical robustness: One view of its use in applications today. *Amer. Statistician* 33, 108-115.

Holland, P.W. & Welsch, R.E. (1977). Robust regression using iteratively reweighted least squares. *Comm. Statist. A*, 813-827.

Huber, P.J. (1981). *Robust Statistics*. Wiley: New York.

Magel, R. & Wright, F.T. (1984). Robust estimates of ordered parameters. *J. Statist. Comput. Simul.* 20, 47-58.

Robertson, T. & Wright, F.T. (1971). A norm reducing property for isotonized Cauchy mean value functions. *Ann. Statist.* 2, 1302-1307.

Sue Leurgans
Department of Statistics
The Ohio State University
1958 Neil Avenue
Columbus, Ohio 43210-1247

AN APPLICATION OF ISOTONIC REGRESSION TO MULTIVARIATE DENSITY ESTIMATION

Thomas W. Sager
University of Texas at Austin

AMS 1980 subject classifications: Primary 62G05, 62H12.
Key words and phrases: Multivariate density estimation, isopleth, isotonic regression, order-restricted inference, dimensionality reduction, consistency, convergence rates.

ABSTRACT

Unrestricted nonparametric multivariate density estimation suffers from difficult convergence and computational problems. One way to overcome these problems is to exploit presumed or estimated structure in the density. The isopleth density estimator presumes or estimates the structure of the contours of the density to effectively reduce the dimensionality. The estimator incorporates an order-preserving algorithm to insure that higher isopleths have higher density estimates than lower isopleths. Convergence properties and a simulation are presented. The importance of edge effects is also noted.

1. **INTRODUCTION.** Applications of isotonic regression sometimes crop up in unexpected places. One such application is described in this paper. Although the techniques of isotonic regression have long been used in the estimation of strongly unimodal densities (Robertson (1967); Wegman (1969, 1970a, 1970b); Sager (1982)), it is both surprising and unappreciated that these techniques can be utilized in the estimation of *any* density, whether unimodal or not, whether univariate or not. What is required is a certain kind of knowledge about the

structure of the density. Not so much knowledge as to pigeonhole the estimation problem into the parametric slot, but not so little as to label it nonparametric. If one can specify or estimate the general form of the isopleths (level curves) of the density, isotonic regression can exploit the order imposed on the density by the isopleths. Just this regimen has been proposed previously (Sager 1986a, 1986b). The motivation of the previous work was to defeat the "curse of dimensionality" (Hand 1982) in density estimation and therefore improve the performance of density estimators in higher dimensions. The convergence rates of conventional density estimators deteriorate badly with increasing dimension (Sager 1986a). Thus, the motif of the previous papers was density estimation *per se*, with isotonic regression playing leitmotif. In this paper we turn our attention to the aspects of that problem which more specifically involve isotonic regression. However, these aspects are inextricably commingled with the density estimation problem. We cannot develop them in isolation. Therefore, we begin with a statement of the problem and a summary of the density estimation results.

2. THE ISOPLETH DENSITY ESTIMATOR. Let $f(x)$ denote a d-dimensional probability density function, $d = 1,2,\cdots$, and $F(x)$ denote its cdf. The statistician's knowledge of the structure of f is critical for the estimation of f. When there is insufficient knowledge to specify a parametric form, there may yet be exploitable information. For example, the statistician may believe that the density is similar to that of a multinormal: unimodal with elliptical contours, but not necessarily multinormal. This knowledge is a description of the contours of the density. Such contours are called isopleths.

DEFINITIONS. An isopleth of f is a set of the form $I_c = \{x; f(x) = c\}$, $0 < c < \infty$. A modal region of f is a set of the form $M_c = \{x; f(x) \geq c\}$. To avoid pathologies, we shall suppose that each modal region is bounded. Also, we shall rule out "flat spots" in the density by assuming that the Lebesgue measure of all isopleths is zero. When f is a composition $f = h(g)$ for some univariate function h, we shall call g an *isopleth form* for f, and h a *transfer density* for f. For example, if f has elliptical contours, then f depends on x only through the value of $g(x) = (x-\theta)'A(x-\theta)$ for some θ and positive definite matrix A; so this g is an isopleth form. Observe that a transfer density h is a true density function on the space of the values of g. Every density function has an isopleth form (for example, the density itself). An isopleth form g is *maximal* if each isopleth of every other isopleth form is contained in an isopleth of g. To ensure maximality, simply choose an isopleth form which assigns a different value to each isopleth of f. Thus we may always redefine a maximal isopleth form to be monotone, in the sense that the value of g on an isopleth is greater than the value of g on any interior isopleth. And the minimum value of g may be chosen to be zero. One convenient way to ensure that g will have these properties is to define $g(x)$ as the volume of the smallest modal region containing x. In fact, the choice of volume for the value of $g(x)$ is so convenient that such a g will be said to be in *standard form*. Many isopleth forms have easily computable volumes. For example, the volume contained in the elliptical modal region $\{(x-\theta)'A(x-\theta) \leq c\}$ is $c^{d/2}\pi^{d/2}(\det A)^{-1/2}/\Gamma(d/2+1)$ (Cramer 1946).

When g is completely specified, the estimation problem is transferred from the (possibly) multivariate f to the univariate h.

Thereby, the curse of dimensionality is laid to rest. But it would be rare for the statistician to know g completely. In the case of an elliptical isopleth form, this would entail providing values for θ and A. More usually, the parameters of the form (e.g., θ and A) would have to be estimated. We first treat the special case in which g is known completely. Later on, we shall see that use of an estimated g is asymptotically equivalent to knowing g completely.

Until further notice, suppose that the isopleth form g is completely known. Let $X_1,\cdots,X_n$ be i.i.d. random variables distributed according to f. Let $X_{(j)}$ be the X_j which yields the $j^{\underline{th}}$ smallest value in $\{g(X_i);\ 1 \le i \le n\}$ and let $C_j = g(X_{(j)})$. It is natural to refer to $X_{(1)},\cdots,X_{(n)}$ as *order statistics*: If $i < j$ then $C_i < C_j$, and $X_{(i)}$ lies higher up on f than does $X_{(j)}$. Moreover, the corresponding modal regions are nested: $M_{C_1} \subset\cdots\subset M_{C_n}$. And any subcollection of $F(M_{C_1}), F(M_{C_2})-F(M_{C_1}),\cdots,F(M_{C_n})-F(M_{C_{n-1}})$ follow a joint Dirichlet distribution, so the areas (or volumes) between modal regions play the role of spacings for order statistics. The essential role played by an isopleth form is thus seen to be that of imposing a natural order in d-dimensional euclidean space. Any x_0 for which an estimated $f(x_0)$ is sought can be thought of as an equivalence class $I_{C_0} = \{x;\ g(x) = g(x_0)\}$, where $C_0 = g(x_0)$. The true density is constant on this equivalence class I_{C_0}; therefore, the estimate should also be constant on the isopleth. Data points near the isopleth convey the same information about the density regardless of whether or not the points are near each other in a euclidean sense.

It now becomes clear how to estimate a multivariate density with specified isopleth form. Let λ denote Lebesgue measure. Since the volume of the regions between successive isopleths plays the role of univariate spacings, we can input the univariate quantities $\lambda(M_{C_1})$, $\cdots,\lambda(M_{C_n})$ as though they were data into *any* univariate density estimator to obtain an estimate $\hat{f}(x) = \hat{h}(g(x))$ of $f(x) = h(g(x))$. (And when g is in standard form, $\lambda(M_{C_i}) = g(X_{(i)})$.) The following theorem shows that when $\hat{f}$ is constructed in this manner, $\hat{f}$ inherits the convergence properties of the univariate estimator $\hat{h}$:

THEOREM 1. *Let* $f = h(g)$ *be a multivariate density function having a known isopleth form* g. *Let* h_n *be a univariate density estimator for* h *based on* $g(X_1),\cdots,g(X_n)$. *For each* x, *let the multivariate density estimate of* f *be* $f_n(x) = h_n(g(x))$. *Then*

(1) f_n *inherits any almost sure convergence rate of* h_n: *if* $a_n(h_n-h) \longrightarrow 0$ *a.s., then* $a_n(f_n-f) \longrightarrow 0$ a.s.

(2) f_n *inherits any distributional limit of* h_n: *if* $a_n(h_n-h) \longrightarrow W$ *in law, then* $a_n(f_n-f) \longrightarrow W$ *in law.*

(3) f_n *inherits any* L_p *limit of* h_n: *if* $a_n\|h_n-h\|_p \longrightarrow 0$ *and the support of* f *is bounded, then* $a_n\|f_n-f\|_p \longrightarrow 0$.

(4) f_n *inherits any mean* L_p *limit of* h_n *(in particular,* MISE): *if* $a_nE(\|h_n-h\|) \longrightarrow 0$ *and the support of* f *is bounded, then* $a_nE(\|f_n-f\|_p) \longrightarrow 0$.

PROOF. See Sager (1986a).

When the assumptions underlying this approach are valid, the problem of high-dimensional density estimation is solved: By exploiting his knowledge of the class of candidate densities, the statistician is able to collapse the problem into essentially a one-dimensional problem, with the computational simplicity and better error properties of the latter. For details, see Sager (1986a, 1986b).

3. **ESTIMATING THE ISOPLETH FORM.** We now suppose that the isopleth form g is not completely known and must be estimated. For example, if the isopleths are known to be elliptical but the parameters θ and A in $g(x) = (x-\theta)'A(x-\theta)$ are unknown, then the natural approach is to estimate θ and A by the sample mean vector and inverse sample covariance matrix, respectively. When substituted into g, these estimates yield an estimate $\hat{g}$ which is consistent for g, subject only to the existence of all second-order moments. In general, it is plausible that if $\hat{g} \longrightarrow g$ and $\hat{h} \longrightarrow h$, then $\hat{h}(\hat{g}) \longrightarrow h(g) = f$, with perhaps some additional side conditions. These expectations are in fact fulfilled, but the argument is a little delicate because $\hat{h}(\hat{g})$ is not a simple composition: the form of $\hat{h}$ depends on its arguments. We can avoid these subtleties by formulating univariate density estimators as functionals T on the space of distribution functions H endowed with the sup norm. Thus $T(H)$ is a density on the real line and $T(H)(z)$ is the value of that density at z.

DEFINITION. We say that a sequence of density estimators T_n is *a.s. consistent for* T if $T_n(H_n)(z) \longrightarrow T(H)(z)$ almost surely, pointwise for all z when $H_n \longrightarrow H$ in sup metric.

DEFINITION. If T_n is a.s. consistent for T, we say that the consistenti is *not-tail-sensitive* if for each z there is a constant $M > |z|$ such that $T_n(H_n^*)(z) \longrightarrow T(H)(z)$ when H_n converges uniformly to H, where

$$H_n^* = \begin{cases} H_n(M) & \text{if } z > M \\ H_n(z) & \text{if } -M \le z \le M. \\ H_n(-M) & \text{if } z < -M \end{cases}$$

The standard density estimators (kernel, nearest neighbor, series, spline, penalized likelihood, isotonic) are not-tail-sensitive under the usual conditions that make them consistent. The principal results which establish the consistency of the isopleth density estimator now follow:

THEOREM 2. *Suppose* $f = h(g)$ *is a bounded multivariate density with isopleth form* g *in standard form and continuous transfer density* h. *Let* $X_1, \cdots, X_n$ *be a random sample distributed as* f. *Let* $\hat{g}(x) = \hat{g}(x;\ X_1, \cdots, X_n)$ *be an estimate of* $g(x)$. *Let* H_n *denote the empirical cdf of the (univariate) quantities* $\hat{g}_1, \cdots, \hat{g}_n$, *where* $\hat{g}_i = \hat{g}(X_i) = \hat{g}(X_i;\ X_1, \cdots, X_n)$. *Let* T_n *be a density estimator, with* $T_n(H_n) = \hat{h}$. *Let* H *denote the cdf of* h. *If*

(1) $\hat{g}(x) \longrightarrow g(x)$ *a.s. for each* x;

(2) H_n *converges to* H *uniformly a.s. on compacta; and*

(3) T_n *is not-tail-sensitive and a.s. consistent for* T, *then* $\hat{h}(\hat{g}(x)) \longrightarrow f(x)$ *a.s. for each* x.

PROOF. See Sager (1986b).

THEOREM 3. *With the same context as Theorem 1, suppose that* g *is parameterized by* υ, *that is,* $g(x) = g(x, \upsilon)$ *where* υ *is a parameter vector. Suppose* $\upsilon_n = \upsilon_n(X_1, \cdots, X_n)$ *is an estimate of* υ *and* $\hat{g}(x) = g(x, \upsilon_n)$. *If*

(1) $g(x, \upsilon)$ *is continuous in* x *and* υ;

(2) $\nu_n \longrightarrow \nu$ *a.s.; and*

(3) T_n *is not-tail-sensitive and a.s. consistent for* T, *then* $\hat{h}(\hat{g}(x)) \longrightarrow f(x)$ *a.s. for each* x.

PROOF. See Sager (1986b).

As a corollary, we immediately have the almost sure consistency of the isopleth density estimator for elliptically contoured densities. We estimate θ and $\mathbf{A}$ by the mean vector and inverse covariance matrix, substitute into $g(\mathbf{x}) = (\mathbf{x}-\theta)'\mathbf{A}(\mathbf{x}-\theta)$, calculate the volumes enclosed by the resulting estimated isopleths $\hat{g}(\mathbf{x}) = (\mathbf{x}-\hat{\theta})'\hat{\mathbf{A}}(\mathbf{x}-\hat{\theta})$, and proceed as though $\hat{g}$ were the true g. As noted above, this will work, subject to finite second-order moments.

4. THE ROLE OF ISOTONIC REGRESSION IN THE ISOPLETH DENSITY ESTIMATOR. Isopleth density estimation is a two-step procedure. The first step is the estimation of the isopleth form $g(\mathbf{x})$. The second is the estiamtion of the density of the isopleth form values. There are at least two ways in which isotonic regression can contribute to this procedure. Both occur in the second of the two steps. First, in the estimation of the univariate transfer density h, no restriction was placed on the choice of density estimator, provided only that it satisfy condition (3) of Theorems 2 and 3 above. With g in standard form (which we henceforth assume without further mention), the transfer density h has its mode at 0 and is monotonically decreasing on its support, which is contained in $[0,\infty)$. Therefore, the isotonic density estimator of Wegman (1969, 1970a, 1970b) is a possible candidate for h. To verify condition (3) for the isotonic density estimator, we note that the pointwise almost sure consistency is given by Theorem 5.1, p. 228, of Barlow, et al. (1972), and the not-tail-sensitivity follows from Lemma B, p. 228 (ibid.). Therefore, by Theorem 1, the isopleth density estimator will inherit the convergence rates of Wegman's isotonic density estimator. In particular, Prakasa Rao (1969) showed that the isotonic density estimator converges in distribution at the rate $n^{-1/3}$. Therefore, so does the isopleth density estimator, when Wegman's estimator is used for $\hat{h}$.

There is a second way in which isotonic regression can assist the isopleth density estimator. Although the true transfer density h is monotone on $[0,\infty)$, we may not wish to estimate h by the isotonic density estimator. We may prefer a nonisotone estimator such as the kernel, spline, nearest neighbor, series, or maximum penalized likelihood density estimators. Since the resulting $\hat{h}$ need not be monotone, we may isotonize the result in order to improve its performance and to confirm it to our knowledge about h. Why might we prefer a nonisotonic estimator for $\hat{h}$? There are at least two reasons. First, the isotonic density estimator converges in distribution at the same rate, $n^{-1/3}$, as the histogram, which is inferior to modern density estimators. Although distributional rates for density estimators depend on the assumptions placed on the true densities and can be forced as close as desired to $n^{-1/2}$ through sufficient restrictions, the usual nonparametric setting results in distributional rates of $n^{-2/5}$. The problem with the histogram results from edge effects with its fixed bins, and the problem with the isotonic estimator is that it does not smooth quite enough data in the construction of its level sets. If the level sets were larger, the variance of the estimator would be less. Secondly, as Wegman (1975) observed, the isotonic density estimator tends to peak too sharply at the mode.

This biasing high at the mode is a direct consequence of characterizations of isotonic (antitonic) regression such as Theorem 1.4, Theorem 7.8, and Definition 7.6 of Barlow et al. (1972). These characterizations show that the isotonic regression loads more mass onto the upper sets around the mode than does the empirical cdf, which is unbiased for the probability content of those upper sets. Consequently, the intervals $[0, g(X_{(i)})]$ containing the first few order statistics (say, $i = 1, \cdots, k$) will receive at least i/n mass. So the estimate $\hat{h}$ will assign values of at least $i/[ng(X_{(i)})]$ to those intervals. For small i, the order statistics $g(X_{(i)})$ are small and have large coefficients of variation. Thus one can expect the isotonic estimator to be unstable and to bias high around the mode and to bias low in the tails. Although the isotonic estimator possesses the maximum (nonparametric) likelihood property, the twin disadvantages of slower convergence rates and modal over-peaking prompt one to seek improvements. One suggestion offered by Wegman (1970a) is to force the first level set around the mode to have at least a prespecified size ϵ. This prevents gross overpeaking prior to applying isotonization. It is in the spirit of our suggestion above to utilize a two-step process in estimating h by first applying a modern density estimator like the kernel or nearest neighbor and then isotonizing the result. Barlow and Van Zwet (1969, 1970) capitalized on a similar idea for generalized failure rate functions (see section 5.4 of Barlow et al. (1972)) by grouping on a grid and then isotonizing, to the betterment of law limit rates. Wright (1982) grouped data prior to isotonizing and improved the convergence rate in estimating a regression function.

The discussion suggests that the two uses of isotonic regression differ with respect to the timing of isotonization. In the first use, we apply the isotonic density estimator directly to the (estimated) isopleth values $g(X_i)$, $i = 1,\cdots,n$. In the second, we first apply a nonisotonic density estimator to the isopleth values and then isotonize the result. The discussion also implies that convergence rates for the first use are inferior to the second. To improve rates for the first use and simultaneously ameliorate its tendency to overpeak, we may consider smoothing it further by applying a density estimator *after* isotonization. With this last idea we complete the symmetry of the two uses: Our choices are (1) to isotonize and then apply a density estimator, or (2) to apply a density estimator and then isotonize. With (2), the result is obviously monotone. With (1), it may not be. However, there are special cases in which (1) will be monotone. We consider one example: the kernel estimator with fixed window width, which is the most commonly used density estimator. The usual kernel estimator is obtained as the density of $X+Y$, where X and Y are independent random variables with X distributed as the empirical cdf and Y an absolutely continuous random variable used to "smooth" X. In the use we envision, the role of X will be played by a random variable distributed as the isotonic density estimator $\hat{h}$. Let k be any kernel density and let $\hat{m}$ be the result of applying k to $\hat{h}$. That is, $\hat{m}(y) = \int\hat{h}(u)k(y-u)du = \int\hat{h}(y-u)k(u)du$. Note that the integrand is 0 for $y < u$. So there are likely to be edge effects for y's close to 0 because the transfer density h has a large discontinuity at 0. In computer simulations, these effects show themselves as large downward biases, which cannot be ignored. One solution is to define $h(u) = h(0)$ for $u \leq 0$. It then follows that $\hat{m}$ is isotonic. For if $y_1 < y_2$, then $\hat{h}(y_1-u) \geq \hat{h}(y_2-u)$ for all u, so that $\hat{m}(y_1) = \int\hat{h}(y_1-u)k(u)du \geq \int\hat{h}(y_2-u)k(u)du = \hat{m}(y_2)$. Another solution is to define $\hat{h}(-u) = \hat{h}(u)$ for all u. Thus, $\hat{h}$ is symmetric about 0. If the kernel k is also assumed to be symmetric and (strongly) unimodal, then $\hat{m}(y) = \int \hat{h}(u)k(y-u)du$ is (strongly) unimodal—that is, monotone—for the convolution of symmetric, unimodal densities is unimodal (Feller (1966), p. 164). In general, however, it is not necessarily true that the convolution of unimodal densities is unimodal. But in most applications, the kernel used is symmetric and unimodal. The author's simulations suggest that either solution corrects the alluded downward biases at 0 but has little effect on the total mass of the estimated density. For alternative approaches to edge effects in density estimation, see Falk (1984) and Swanepoel and Van Wyk (1981a, 1981b).

Finally, we examine the effect of the timing of isotonization on the convergence rate of $\hat{h}$. We once again formulate density estimators as functionals on the space of distribution functions. Let I denote the isotonic density estimator fucntional and let D denote another density estimator functional with additional properties to be specified later. We denote by $\bar{I}$ or $\bar{D}$ the distribution function corresponding to the image of I or D. With this notation we can represent our choice between the two uses of isotonic regression in isopleth density estimation as the choice between computing (1) $\hat{h} = D(\bar{I}(H_n))$ or (2) $\hat{h} = I(\bar{D}(H_n))$, where H_n denotes the empirical cdf of the $g(X_i)$, $i = 1,\cdots,n$. We say that D preserves monotonicity if D maps distribution functions of monotone densities into monotone densities. Our result on convergence rates is contained in the following theorem, which applies to the general estimation of monotone densities—not just in the isopleth context. It says that isotonization does not harm the almost sure convergence rate of the density estimator, even though the isotonization procedure alone may have a slower rate.

THEOREM 4. *Let* Q_n *denote the empiric cdf of a random sample from a distribution having monotone density* q *and cdf* Q.

(i) *Let* D *be a density estimator functional based on a symmetric unimodal fixed-width kernel* k. *Let* $a_n = o((n/\log\log n)^{1/2})$. *If* $a_n|\overline{D}(Q_n)(x)-Q(x)| \to 0$ *almost surely for each* x *and* $a_n|D(Q_n)(x)-q(x)| \to 0$ *almost surely for each* x, *then* $a_n|D(\overline{I}(Q_n))(x)-q(x)| \longrightarrow 0$ *almost surely for each* x, *where we assume* $I(Q_n)$ *is symmetrized prior to applying* D.

(ii) *Let* D *be a density estimator functional which maps into the set of continuous densities. If* $a_n \sup_x|D(Q_n)(x)-q(x)| \longrightarrow 0$ *almost surely, then* $a_n \sup_x|I(\overline{D}(Q_n))(x)-q(x)| \longrightarrow 0$ *almost surely.*

PROOF. (i) We first establish the claim that $a_n|\overline{D}(\overline{I}(Q_n))(x)-Q(x)| \longrightarrow 0$ almost surely for each x. To show this, observe that $\overline{D}(\overline{I}(Q_n))(x) = \int_{-\infty}^{x}\int_{-\infty}^{\infty} I(Q_n)(u)k(y-u)\,du\,dy = \int_{-\infty}^{\infty} k(u)\int_{-\infty}^{x} I(Q_n)(y-u)\,dy\,du = \int_{-\infty}^{\infty}k(u)\overline{I}(Q_n)(x-u)\,du$. But by Lemma A, p. 227 of Barlow et al. (1972), $\overline{I}(Q_n)(x) = Q_n(x) + O(\sup_z|Q_n(z)-Q(z)|)$. It is well known that the latter is $Q_n(x) + O((\log\log n/n)^{1/2})$ (for example, Chung (1949)). So

$$
\begin{aligned}
a_n|\overline{D}(\overline{I}(Q_n))(x)-Q(x)| \\
&= a_n\left|\int_{-\infty}^{\infty}k(u)\{Q_n(x-u)+O((\log\log n/n)^{1/2})\}du-Q(x)\right| \\
&= a_n\left|\int_{-\infty}^{\infty}k(u)Q_n(x-u)du-Q(x)\right| + a_n\cdot O((\log\log n/n)^{1/2}) \\
&= a_n|\overline{D}(Q_n)(x)-Q(x)| + a_n\cdot O((\log\log n/n)^{1/2}) \longrightarrow 0
\end{aligned}
$$

by the hypotheses.

With the claim established, we next observe that D preserves monotonicity (see discussion preceding Theorem 4). Thus the cdf $\overline{D}(\overline{I}(Q_n))$ is concave. Thus for $\epsilon > 0$, the geometry of concavity yields

$$[a_n\overline{D}(\overline{I}(Q_n))(x)-a_n\overline{D}(\overline{I}(Q_n))(x-\epsilon)]/\epsilon \geq a_nD(\overline{I}(Q_n))(x)$$
$$\geq [a_n\overline{D}(\overline{I}(Q_n))(x+\epsilon)-a_n\overline{D}(\overline{I}(Q_n))(x)]/\epsilon.$$

Thus,

$$\{a_n[\overline{D}(\overline{I}(Q_n))(x)-Q(x)] - a_n[\overline{D}(\overline{I}(Q_n))(x-\epsilon)-Q(x-\epsilon)]\}/\epsilon$$
$$\geq a_n\{D(\overline{I}(Q_n))(x)-[Q(x)-Q(x-\epsilon)]/\epsilon\},$$

and

$$a_n\{D(\overline{I}(Q_n))(x)-[Q(x+\epsilon)-Q(x)]/\epsilon\}$$
$$\geq \{a_n[\overline{D}(\overline{I}(Q_n))(x+\epsilon)-Q(x+\epsilon)]-a_n[\overline{D}(\overline{I}(Q_n))(x)-Q(x)]\}/\epsilon.$$

Let $n \longrightarrow \infty$ and apply the claim we just established to get

$$0 \geq \lim a_n\{D(\overline{I}(Q_n))(x)-[Q(x)-Q(x-\epsilon)]/\epsilon\}$$

and

$$\lim a_n\{D(\overline{I}(Q_n))(x)-[Q(x+\epsilon)-Q(x)]/\epsilon\} \geq 0.$$

Since this is true for all $\epsilon > 0$, we have $\lim a_n[D(\overline{I}(Q_n))(x)-q(x)] = 0$.

(ii) We show that $\sup_x a_n|I(\overline{D}(Q_n))(x)-q(x)| \leq \sup_x a_n|D(Q_n)(x) -q(x)|$. Now $I(\overline{D}(Q_n))$ is the slope from the left of the least concave majorant of $\overline{D}(Q_n)$. So if there is a point x_0 at which $I(\overline{D}(Q_n))$ and $D(Q_n)$ do not coincide, then there is an interval

$(a,b]$, with $a < x_0 < b$ and $I(\overline{D}(Q_n))$ constant on $(a,b]$, $\overline{I}(\overline{D}(Q_n))(a) = \overline{D}(Q_n)(a)$, $\overline{I}(\overline{D}(Q_n))(b) = \overline{D}(Q_n)(b)$, and $\overline{I}(\overline{D}(Q_n))(x) > \overline{D}(Q_n)(x)$ for $a < x < b$. There are three cases:

(a) $D(Q_n)(x_0) \geq \overline{I}(D(Q_n))(x_0) \geq q(x_0)$;

(b) $D(Q_n)(x_0) \geq q(x_0) \geq I(\overline{D}(Q_n))(x_0)$; and

(c) $q(x_0) \geq D(Q_n)(x_0) \geq I(\overline{D}(Q_n))(x_0)$.

In case (a), we immediately have $|I(\overline{D}(Q_n))(x_0)-q(x_0)| \leq |D(Q_n)(x_0) -q(x_0)|$. In cases (b) and (c), the constancy of $I(\overline{D}(Q_n))$ and the monotonicity of q show that $|I(\overline{D}(Q_n))(x)-q(x)|$ increases as $x \longrightarrow a$; since $\overline{I}(\overline{D}(Q_n))$ lies above $\overline{D}(Q_n)$, the continuity of $D(Q_n)$ and the mean value theorem guarantee the existence of an x_1 near a such that

$$q(x_1) \geq I(\overline{D}(Q_n))(x_1) \geq D(Q_n)(x_1) = \{\overline{D}(Q_n)(x_1)-\overline{D}(Q_n)(a)\}/(x_1-a).$$

Thus,

$$|D(Q_n)(x_1)-q(x_1)| \geq |I(\overline{D}(Q_n))(x_1)-q(x_1)| \geq |I(\overline{D}(Q_n))(x_0)-q(x_0)|.$$

The result follows.

5. AN EXAMPLE. To illustrate the isopleth density estimator, a simulation was performed. One hundred independent observations $X_1,\cdots,X_{100}$ were generated with common bivariate normal (μ,V) distribution, where $\mu = 0$ and $V = \{1,.7/.7,1\}$. The normal density depends on x only through the value of $(x-\mu)'V^{-1}(x-\mu)$, which is an isopleth form for the normal density. The area contained in the elliptical region $\{x;\ (x-\mu)'V^{-1}(x-\mu) \leq c\}$ is $c\pi/(\det V)^{-1/2}$ (cf. Cramer (1946)). Thus, when written in standard form so that $g(x) =$ area enclosed by the isopleth containing x, the true isopleth form is $g(x) = (x-0)'V^{-1}(x-0)\pi(\det V)^{-1/2} = (1.96x_1^2-2.745x_1x_2+1.96x_2^2)\pi(.71414)$

$= 4.3973x_1^2 - 6.1585x_1x_2 + 4.3973x_2^2$. The simulated data yielded $\mu = (0.05434, 0.06086)$ and $V^{-1} = \{1.9331, -1.3519/-1.3519, 1.9336\}$. Thus the estimated isopleth form was $\hat{g}(x) = 4.3939x_1^2 - 6.1453x_1x_2 + 4.3950x_2^2 - .1035x_1 - .2010x_2 + 0.008929$.

The first step in computing the isopleth density estimate is to estimate the isopleth form. This has just been done. The second step is to estimate the transfer density. To this end, the g-values for each of the 100 observations were computed by substituting $X_1, \cdots, x_{100}$ into the estimated $\hat{g}$, above. Then the density of the $\hat{g}$-values was estimated by treating the $\hat{g}$'s as a random sample and applying a univariate density estimator. The estimator chosen was a $k^{\underline{th}}$ nearest neighbor density estimator with $k = 10$. The resulting nearest neighbor density $\tilde{h}$ was not monotone, so it was isotonized as suggested in section 4 by the pool-adjacent-violators algorithm to yield the estimated transfer density $\hat{h}$. Table 1 displays the values, and Figure 1 shows a plot of the estimated and true transfer densities. The true transfer density is that of π times a chi-square variable with two degrees of freedom. Finally, the isopleth density estimate is plotted (or computed) for each desired x by "looking up" the value of the transfer density associated with $\hat{g}(x)$. Figure 2 is the finished product. The level sets of the isotonic regression are clearly visible as flat spots in the figure. A smoother figure could be obtained, if desired, by choosing a larger value for k in the nearest neighbor algorithm or by interpolating continuously over the level sets.

Table 1. 100 simulated i.i.d. N(0,0,1,1,.7), data (X1,X2) area of enclosed modal region (G), nearest neighbor (k=10), density estimate (HN), isotonized density estimate (HNISO).

OBS	X1	X2	G	HN	HNISO
1	0.1588	0.1206	0.02528	0.212196	0.214648
2	-0.0461	0.0530	0.03977	0.212196	0.214648
3	0.0387	0.1943	0.09210	0.212196	0.214648
4	-0.2473	-0.1464	0.20441	0.212196	0.214648
5	-0.2587	-0.1549	0.22008	0.212196	0.214648
6	0.2838	0.3745	0.22266	0.212196	0.214648
7	-0.2566	-0.1954	0.22373	0.211150	0.214648
8	0.2557	0.4243	0.30887	0.221612	0.214648
9	-0.3309	-0.0184	0.49200	0.184810	0.214648
10	0.4892	0.5236	0.53536	0.176592	0.214648
11	0.1161	0.4530	0.54367	0.174782	0.214648
12	-0.0637	0.3255	0.56073	0.157673	0.214648
13	-0.2167	0.1817	0.58847	0.165871	0.214648
14	0.6500	0.4518	0.79962	0.226451	0.214648
15	-0.2884	-0.5418	0.187632	0.187632	0.214648
16	0.4314	-0.0236	0.85202	0.164831	0.214648
17	0.3791	-0.1068	0.92138	0.167694	0.214649
18	0.5437	0.0887	0.97203	0.149181	0.214648
19	0.6075	0.1904	0.97776	0.208971	0.214648
20	0.0851	-0.4223	1.12161	0.227139	0.214648
21	0.1897	0.6714	1.21101	0.210693	0.214648
22	-0.6196	-0.1987	1.21669	0.238229	0.214648
23	-0.4413	-0.7054	1.32583	0.244337	0.214648
24	0.0525	-0.4897	1.32601	0.239133	0.214648
25	0.5952	0.8295	1.32728	0.326938	0.214648
26	-0.7228	-0.5463	1.37411	0.242829	0.214648
27	-0.2863	-0.6827	1.38312	0.240165	0.214648
28	-0.5303	-0.7338	1.42223	0.289022	0.214648
29	-0.0388	0.5638	1.43775	0.260936	0.214648
30	0.1689	0.71111	1.45807	0.230846	0.214648
31	0.3757	-0.2853	1.66401	0.226303	0.214648
32	0.8953	0.5083	1.67471	0.221729	0.214648
33	0.8612	0.8608	1.70642	0.179365	0.179365
34	-0.0419	0.6204	1.74757	0.128540	0.147918
35	0.9369	0.5664	1.80378	0.120524	0.147918
36	-0.1596	0.5436	1.86018	0.135429	0.147918
37	-0.2139	0.4985	1.87922	0.134204	0.147918
38	0.4062	0.9392	2.03550	0.139585	0.147918
39	0.9676	0.3900	2.29352	0.145484	0.147918
40	-0.6433	-0.9661	2.37075	0.150675	0.147918
41	0.8051	1.1106	2.47624	0.157459	0.146918
42	0.8319	0.0959	2.49435	0.155757	0.147918
43	0.4384	1.0310	2.49447	0.148345	0.147918
44	0.8631	0.1411	2.50366	0.165776	0.147918
45	0.6313	1.1018	2.53383	0.157630	0.147918
46	-1.0011	-0.7911	2.55877	0.166427	0.147918
47	-0.7923	-0.0598	2.58545	0.165014	0.147918
48	-0.8251	-0.0675	2.77702	0.105390	0.122742
49	-0.8961	-1.0633	2.95706	0.102746	0.122742
50	-1.0784	-0.5246	3.06859	0.104009	0.122742

Table 1. Continued.

OBS	X1	X2	G	HN	HNISO
51	-0.2773	0.6399	3.13719	0.102827	0.122742
52	-0.5447	-1.0900	3.16096	0.104399	0.122742
53	1.2520	1.1668	3.53822	0.109928	0.122742
54	-1.0951	-1.1144	3.57427	0.117006	0.122742
55	-0.7062	-1.1931	3.59143	0.127737	0.122742
56	0.1675	1.0450	3.62854	0.122997	0.122742
57	-1.1169	-0.4015	3.63911	0.115199	0.122742
58	1.3028	1.1850	3.77768	0.181690	0.122742
59	-1.0303	-1.2319	3.8972	0.178972	0.122742
60	1.1146	0.2366	3.9297	0.108855	0.108855
61	-0.4119	0.6328	4.0315	0.103578	0.103578
62	1.3635	1.2227	4.1158	0.093434	0.099014
63	1.3948	0.8409	4.1436	0.104594	0.099014
64	1.2851	1.3446	4.1889	0.084186	0.085652
65	1.4635	1.2262	4.6020	0.083470	0.085652
66	1.3681	1.4099	4.6905	0.089302	0.089302
67	1.5188	1.0786	4.8164	0.073335	0.073335
68	-0.5968	0.5449	4.8294	0.060667	0.060667
69	-0.5095	0.6774	5.2038	0.052071	0.058487
70	0.1870	1.2422	5.2476	0.059208	0.058487
71	-1.1749	-0.1463	5.2633	0.061276	0.058487
72	-1.0970	-1.5190	5.6158	0.061394	0.058487
73	1.6826	1.2265	5.9568	0.053813	0.054047
74	1.2653	0.0801	6.3014	0.054025	0.054047
75	1.1743	1.7543	6.4598	0.054303	0.054047
76	-1.0383	-1.6335	6.4857	0.046503	0.048344
77	-1.6157	-0.8266	6.6081	0.045713	0.048344
78	-0.9667	0.3683	6.8735	0.052815	0.048344
79	1.5231	1.8260	7.2399	0.038326	0.038326
80	-1.6591	-0.7437	7.2733	0.033992	0.034024
81	0.0695	-0.2085	7.6287	0.032567	0.034024
82	0.6052	-0.8464	8.0221	0.032909	0.034024
83	-1.8373	-1.2927	8.0396	0.034016	0.034024
84	-1.2894	-1.9576	9.1715	0.035490	0.034024
85	-0.2922	1.2831	9.6985	0.033089	0.034024
86	1.5604	0.0720	9.8633	0.035792	0.034024
87	2.1371	1.2990	9.9506	0.031670	0.034024
88	-1.3483	-2.0575	10.1072	0.029995	0.034024
89	1.8832	0.5372	10.3393	0.040541	0.034024
90	-1.6946	-2.0830	10.5967	0.029287	0.029287
91	-0.8289	0.8692	10.6867	0.023035	0.023035
92	0.4563	1.9334	11.4955	0.022474	0.022474
93	2.1063	2.2112	11.7068	0.020840	0.020840
94	-2.0892	-2.0348	11.8848	0.018738	0.018738
95	1.6849	-0.1038	13.4518	0.007367	0.007403
96	-1.4601	-2.4672	14.6387	0.007411	0.007403
97	-1.2854	-2.4444	14.8452	0.007411	0.007403
98	-2.5292	-1.4450	15.3855	0.007411	0.007403
99	0.5154	-1.5088	16.2098	0.007411	0.007403
100	-1.2449	1.3766	25.5299	0.007411	0.007403

Figure 1. Estimated (+) and True (Smooth) Transfer Densities
100 Simulated Bivariate Normal (0, 0, 1, 1, .7) Data

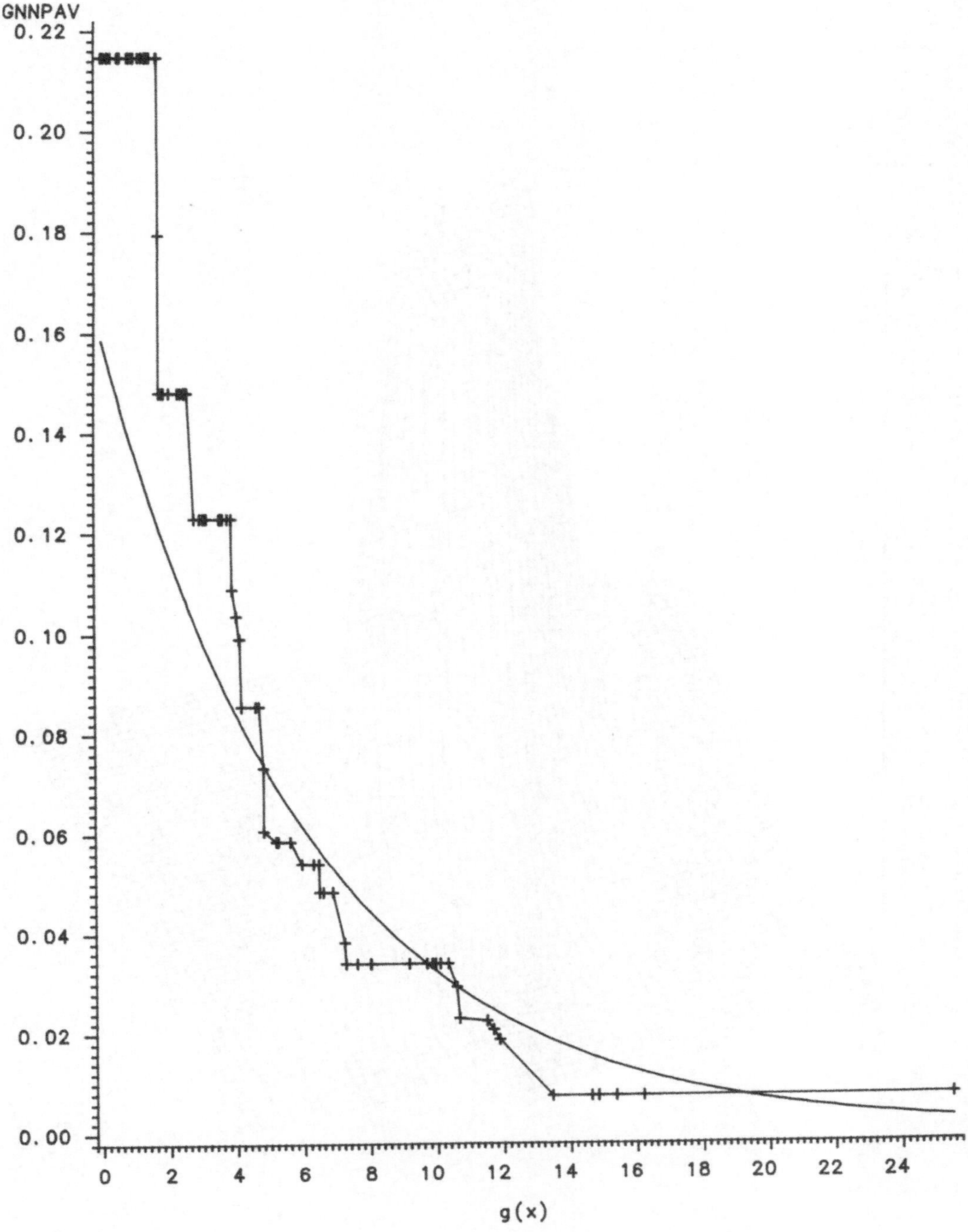

Figure 2. Isopleth Density Estimate
100 Simulated Bivariate Normal (0, 0, 1, 1, .7) Data

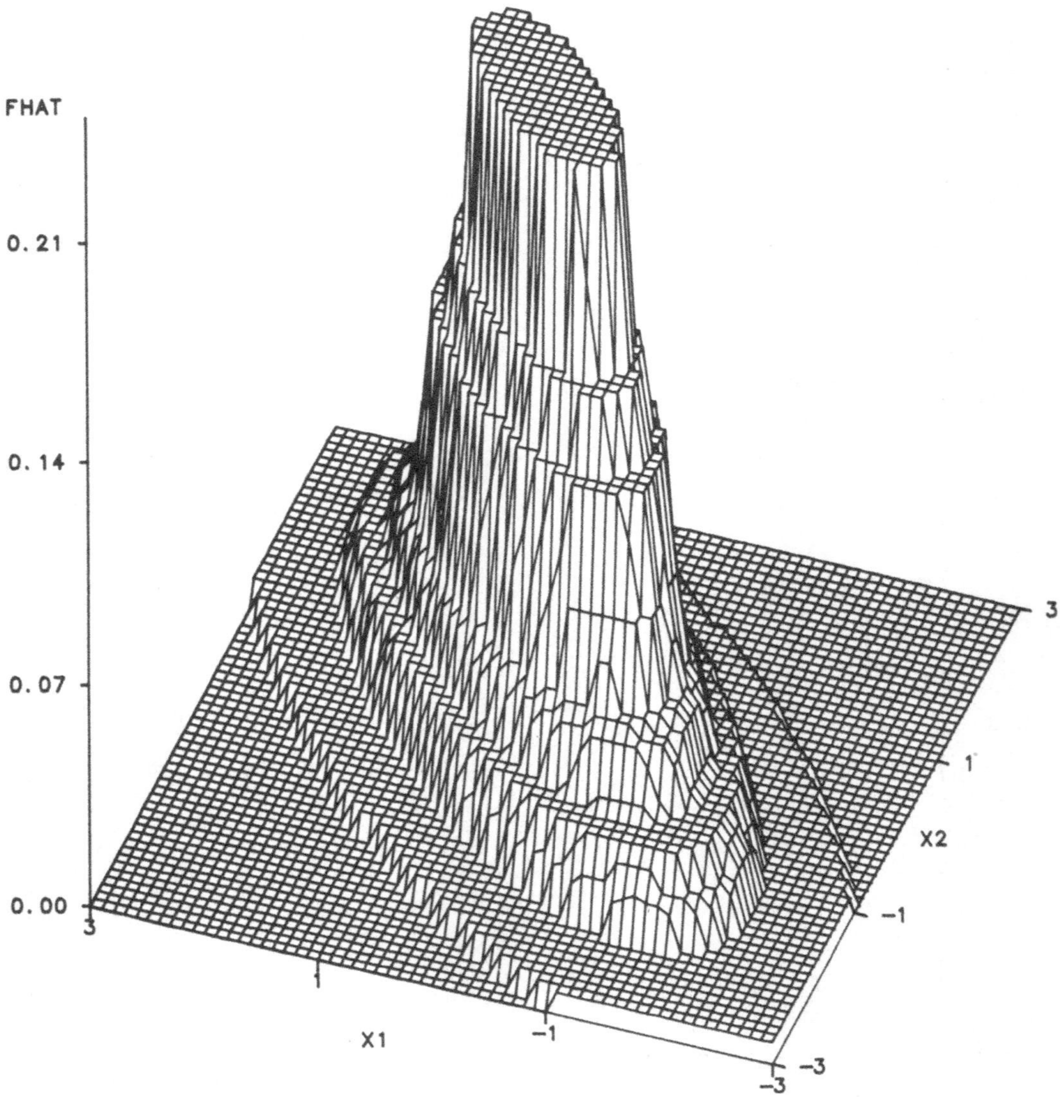

REFERENCES

Barlow, R.E., Bartholomew, D.J., Bremner, J.M. & Brunk, H.D. (1972). *Statistical Inference Under Order Restrictions*. Wiley, New York.

Barlow, R.E. & Van Zwet, W.R. (1969). Asymptotic properties of isotonic estimators for the generalized failure rate function. Part II: asymptotic distributions. Operations Research Center Report ORC 69-10, University of California, Berkeley.

Barlow, R.E. & Van Zwet, W.R. (1970). Asymptotic properties of isotonic estimators for the generalized failure rate function. Part I: strong consistency. *Nonparametric Techniques in Statistical Inference*, M.L. Puri (Ed.). Cambridge University Press, 159-173.

Chung, K.L. (1949). An estimate concerning the Kolmogoroff limit distribution. *Trans. Amer. Math. Soc.* 67, 36-50.

Cramer, H. (1946). *Mathematical Methods of Statistics*. Princeton University Press, Princeton, New Jersey.

Falk, M. (1984). Kernel estimation of a dsensity in an unknown endpoint of its support. *South African Statist. J.* **18**, 91-96.

Feller, W. (1966). An Introduction to Probability Theory and Its Applications, Vol. II. New York: Wiley.

Hand, D.J. (1982). *Kernel Discriminant Analysis*. New York: Wiley.

Prakasa Rao, B.L.S. (1969). Estimation of a unimodal density. *Sankhya Ser. A*, **31**, 23-26.

Robertson, T.J. (1967). On estimating a density which is measurable with respect to a σ-lattice. *Ann. Math. Statist.* **38**, 482-493.

Sager, T.W. (1982). Nonparametric maximum likelihood estimation of spatial patterns. *Ann. Statist.* **10**, 1125-1136.

Sager, T.W. (1986a). Some isopleth methods for mapping multidimensional distributions. In *Statistical Image Processing and Graphics*, E.J. Wegman (Ed.). Marcel Dekker (to appear).

Sager, T.W. (1986b). Searching for and exploiting structure in higher dimensional density estimation. In *Classification as a Tool of Research*, W. Gaul & M. Schader (Eds.). North-Holland, 389-396.

Swanepoel, J.W.L. & Van Wyk, J.W.J. (1981a). Fixed width confidence intervals for the truncation parameter of an unknown distribution function. *South African Statist. J.* 15, 161-166.

Swanepoel, J.W.L. & Van Wyk,J.W.J. (1981b). Nonparametric estimation of the derivative of an unknown distribution function in some unknown truncation point. *South African Statist. J.* 15, 167-172.

Swanepoel, J.W.L. & Van Wyk, J.W.J. (1981b). Nonparametric estimation of the derivative of an unknown distribution function in some unknown truncation point. *South African Statist. J.* 15, 167-172.

Wegman, E.J. (1969). A note on estimating a unimodal density. *Ann. Math. Statist.* **40**, 1661-1667.

Wegman, E.J. (1970a). Maximum likelihood estimate of a unimodal density function. *Ann. Math. Statist.* **41**, 457-471.

Wegman, E.J. (1970b). Maximum likelihood estimation of a unimodal density, II. *Ann. Math. Statist.* **41**, 2169-2174.

Wegman, E.J. (1975). Maximum likelihood estimation of a probability density function. *Sankhya Ser. A*, **37**, 211-224.

Wright, F.T. (1982). Monotone regression estimates for grouped observations. *Ann. Statist.* **10**, 278-286.

Thomas W. Sager
Department of General Business
University of Texas at Austin
Austin, TX 78712

LEAST-SQUARES REGRESSION UNDER CONVEXITY AND HIGHER-ORDER DIFFERENCE CONSTRAINTS WITH APPLICATION TO SOFTWARE RELIABIILTY[1]

Douglas R. Miller
The George Washington University

Ariela Sofer
George Mason University

AMS 1980 subject classifications: 60K10, 62M99, 62N05, 65K05, 68N99, 90C20, 90C50.

Keywords: *constrained inference, quadratic programming, software reliability, complete monotonicity.*

ABSTRACT

The isotone regression problem of finding a least squares isotone sequence is extended by imposing order restrictions also on higher order differences of the sequence. This new problem has a number of applications, and one example in the area of software reliability is presented.

In contrast to the isotone regression problem, there is no simple finite algorithm for solving least squares problems when higher order differences are also order restricted. The paper discusses some of the numerical difficulties which may arise due to the ill-posed nature of the problem and outlines a numerically stable algorithm for solving it.

---

1. Research supported by National Aeronautics and Space Administration Grant NAG-1-179.

1. INTRODUCTION. Consider the following problem:

$$(*) \quad \underset{x_1,\cdots,x_r}{\text{minimize}} \sum_{i=1}^{n} w_i(x_i-r_i)^2$$

$$(1.1) \qquad \text{subject to} \quad (-1)^j \Delta^j x_i \geq 0, \quad j+1 \leq i, \quad j = 0,1,\cdots,d$$

where w_i $(i = 1,\cdots,n)$ are positive weights, and Δ^j is the $j^{\underline{th}}$ order backward difference operator, i.e.,

$$\Delta^0 x_i = x_i$$

$$\Delta^1 x_i = x_i - x_{i-1}$$

$$\Delta^j x_i = \Delta^{j-1}x_i - \Delta^{j-1}x_{i-1}, \quad i < j.$$

The problem therefore is that of least squares regression under higher order difference constraints. For $d = 1$ this yields the well known isotone (antitone) regression problem; see McWilliams (1979) for an investigation of this case. For $d = 2$ this is the problem of least squares antitone regression under convexity restrictions. Our own interest in this problem was motivated by an application in software reliability, for which the maximal order of difference d is 2 or greater. In this paper we present this application and report on our computational experience in solving the associated problem (*).

Prior to solving (*) it behooves us to simplify the system of inequalities (1.1). Using the fact that

$$(-1)^{j-1}\Delta^{j-1}x_{i-1} + (-1)^j\Delta^j x_i = (-1)^{j-1}\Delta^{j-1}x_{i-1},$$

it is simple to show inductively that the constraints of (1.1) for which $j < d$ and $i < n$ are redundant. Thus, for example, the monotonicity constraints $\Delta^1 x_i \le 0$, $i = 2,\cdots,n-1$ are implied by the constraints $\Delta^1 x_n \le 0$ and $\Delta^2 x_i \ge 0$, and hence these constraints need not be posed explicitly. It follows that the system of inequalities (1.1) is equivalent to the reduced system

$$(1.2) \qquad \begin{aligned} &(-1)^d \Delta^d x_i \ge 0, && d+1 \le i \le n \\ &(-1)^j \Delta^j x_n \ge 0, && 0 \le j \le d-1 \end{aligned}$$

The system above is a linear system of the form $Ax \ge 0$ where the $n \times n$ upper triangular matrix A is defined by

$$(1.3) \qquad a_{ij} = \begin{cases} (-1)^{j-i} \binom{d}{j-i} & \begin{array}{l} i \le j \le i+d \\ 1 \le i \le n-d \end{array} \\ (-1)^{j-i} \binom{n-i}{j-i} & \begin{array}{l} i \le j \le n \\ n-d+1 \le i \le n \end{array} \\ 0 & \text{otherwise.} \end{cases}$$

Let $r = (r_1,\cdots,r_n)^T$, $W = \text{diag}(w_1,\cdots,w_n)$. Then the optimization problem can be written as

$$\begin{aligned} &\underset{x \in R^n}{\text{minimize}} \quad 1/2(x-r)^T W(x-r) \\ &\text{subject to} \quad Ax \ge 0. \end{aligned}$$

In contrast to isotone regression, where the pool adjacent violators algorithm (see Barlow, Bartholomew, Bremner and Brunk (1972)) solves the problem in a finite number of simple steps, there is no simple finite algorithm for solving restricted least squares problems when the additional constraints of isotone (or antitone) higher order differences are imposed. In order to solve this problem we tested a number of different solution methods. This paper does not by any means constitute a comprehensive study of solution methods. Rather, the intent of the paper is to point out some of the numerical difficulties which may arise in the solution process. These difficulties, which include problems of sensitivity to round off errors and nonconvergence, are due to the ill conditioning of the matrix A. A discussion of this ill conditioning is given in Appendix A. As a result of this ill conditioning, it is essential that any algorithm used to solve our problem be numerically stable. A description of one such algorithm is given in Section 3.

The method we tested to solve the problem can broadly be classified under two different approaches: (a) methods which move through a sequence of feasible points to the optimal solution; (b) methods which converge to the solution from the exterior of the feasible region.

Methods which use the first approach are sometimes referred to as primal methods. At each iteration of such methods, a movement is made from the current feasible point to a new better point (i.e., a point with a lower objective value). One advantage of this approach is that even if the program is terminated, a feasible suboptimal solution will still be available. The specific algorithms we tested are based on active set methods. Active set methods have been widely discussed in the mathematical programming literature. Detailed expositions can be found in references such as McCormick (1983) and Gill and Murray (1978). These methods solve a sequence of problems that minimize the objective with a subset of the inequalities fixed as equalities.

Though each iteration is fairly complex, the number of iterations is finite. In Section 3, we give a brief description of these methods and present two different versions which we tested. Although these two versions differ only by the numerical algebra implementation, one is by far superior to the other in cases of ill conditioning.

As an alternative, we also coded Dykstra's algorithm (1983) for restricted least squares regression. The sequence of points generated by this method is infeasible — at each iteration (except, in the case of convergence, the last) the current point violates some of the constraints. However, the sequence of maximal constraint violations generated by this iterative method converges to zero. Theoretically, the number of iterations required for convergence could be infinite. Therefore, in practice, the program is terminated when the maximal constraint violation is smaller than some prescribed tolerance ϵ (that is, when the problem constraints are satisfied to within ϵ). Thus, the number of iterations required should be finite.

The layout of the paper is as follows: Section 2 presents an application in software reliability of least squares regression under high order difference constraints. In Section 3 we give a brief overview of Newton-type active set methods, and consider two numerical inplementations. While one of these two versions is computationally fast, it may be numerically unstable. The other version, which uses orthogonal matrix factorizations, requires more computational work but is numerically stable. Due to the ill conditioned nature of the problem, it is this latter method which we recommend using. Section 4 gives a discussion of the performance of the algorithms, with emphasis on the numerical difficulties which may arise when solving the problem.

2. REGRESSION ESTIMATES OF SOFTWARE RELIABIILTY GROWTH. Computer programs usually contain design faults, called "bugs." A usual practice consists of using or executing a program until a bug manifests itself as a failure, then correcting the code so as to remove the bug, and then continuing use or execution. If bugs are corrected without introducing new faults, the program evolves into a more reliable program; hence the term "reliability growth." This debugging process may be a last stage in software development or may also take place in the usage environment after release of the product. In either case, people are interested in the reliability of the software. We can make inferences about the reliability from the debugging data.

Suppose a program is executed for a length of time T. During this time, n bugs are detected and removed when they manifested themselves as failures; these failure times are: $0 \le t_1 \le t_2 \le \cdots \le t_n \le T$. At time T, there are several questions concerning the program: (i) What is the expected time until the next bug occurs? (ii) What is the current failure rate of the program? (iii) How many bugs do we expect to see over some finite future horizon? (iv) How much more debugging is required until some target reliability is obtained? It is possible to make some inferences along these lines using the data $\{t_1, t_2, \cdots, t_n\}$; the usual approach is to fit a parametric model to the data. We shall consider a nonparametric approach. (Miller and Sofer (1985) previously introduced this statistical approach using the projection matrix implementation of the active set method discussed in Section 3.)

We focus here on the failure rate of the program. Let $N(t)$ equal the number of bugs observed in $[0,t]$ and $M(t) = EN(t)$ the expected number. We define the failure rate as $r(t) = dM(t)/dt$. The failure rate function $r(\cdot)$ can be estimated from the data by using

$\{N(t), 0 \leq t \leq T\}$ as an estimate of $M(t)$ and then doing some smoothing to get a smooth function; the slope of this function is an estimate of $r(t)$, $0 \leq t \leq T$. Raw estimates of the failure rate are typically very noisy and of little use; see Miller and Sofer (1985) for an example. We propose estimating as follows: first we estimate $M(t)$, $0 \leq t \leq T$, with a continuous function

$$(2.1) \qquad \hat{M}(t) = \begin{cases} i+(t-t_i)/(t_{i+1}-t_i), & t_i \leq t \leq t_{i+1}, \quad i = 0,1,\cdots,n-1 \\ n+.5(t-t_n)/(T-t_n), & t_n \leq t \leq T. \end{cases}$$

The slope of $\hat{M}(\cdot)$ at t is an estimate of $r(t)$. (Note that the factor of .5 in the final interval reflects the absence of a failure at one end of the interval. There exist rationals for and against including this factor. The choice is somewhat arbitrary and should probably be based on the performance of the procedure rather than a priori considerations.) We shall need a discrete estimate of the failure rate: discretize $[0,T]$ into k intervals of equal length, letting $\Delta s = T/k$ and $s_i = i\Delta s$, $i = 0,1,\cdots,k$, and define

$$(2.2) \qquad \hat{r}_i = (\hat{M}(s_i) - \hat{M}(s_{i-1}))/\Delta s, \quad i - 1,2,\cdots,k.$$

This constitutes a raw estimate of the failure rate. An estimate of current reliability is $\hat{r}_k$; however, this is a very noisy estimator. It can be improved by smoothing.

We wish to take a nonparametric approach to estimating $r(\cdot)$, however, the family of admissible estimates must still be restricted. A natural property to require is *complete monotonicity*. A function $r(\cdot)$ on $[0,\infty)$ is completely monotone if it possesses derivatives of all orders and they alternate in sign (Feller (1971)):

(2.3) $$(-1)^n \frac{d^n r(t)}{dt^n} \geq 0, \quad 0 \leq t < \infty, \quad n = 0,1,2,\cdots.$$

There are two main reasons for requiring complete monotonicity: The first reason is that virtually all of the various (and competing) software reliability growth models in the literature have this property; see Jelinsky and Moranda (1972), Littlewood (1981), Goel and Okumoto (1979), Musa and Okumoto (1984), Crow (1974), and Duane (1964). (The complete monotonicity of these models is verified by differentiating the mean function of the process and checking (2.3).) The second reason is that if the times until each bug manifests itself are (nonidentically-distributed) exponential random variables, then by definition of the exponential distribution and the additive property of the expectation operator

(2.4) $$M(t) = \sum_{i=1}^{\infty} (1-e^{-\lambda_i t}), \quad r(t) = \sum_{i=1}^{\infty} \lambda_i e^{-\lambda_i t}.$$

It can be seen that $r(\cdot)$ is the Laplace transform of a measure which puts mass λ_i at $t = \lambda_i$, $i = 1,2,\cdots$, and therefore must be completely monotone, Feller (1971). Furthermore, since every completely monotone function on $[0,\infty)$ is a Laplace transform of a measure on $[0,\infty)$ some deneness-type properties hold and complete monotonicity characterizes admissible rate functions; see Miller (1986) for additional detail and discussion. (Note that these models are much more applicable to systems which are moderately reliable than to systems for which ultrahigh reliability must be verified.)

Thus we wish to find a completely monotone function which in some sense "best" fits the failure data. It appears necessary to discretize the time axis in order to work with sequences rather than functions. The analogue of a completely monotone function is a completely monotone sequence, $\{r_1, r_2, r_3, \cdots, r_k, \cdots\}$. Define the difference operator

$$\Delta^0 r_i = r_i; \quad \Delta^n r_i = \Delta^{n-1} r_i - \Delta^{n-1} r_{i-1}, \quad n = 1, 2, \cdots.$$

The sequence $\{r_1, r_2, \cdots, r_k, \cdots\}$ is completely monotone if

$$(-1)^n \Delta^n r_i \geq 0, \quad i = n+1, n+2, \cdots, \quad n = 0, 1, 2, \cdots. \tag{2.5}$$

Our goal is to find the "best-fitting" completely monotone sequence to $\{r_1, r_2, \cdots, r_k\}$ and use it as an estimate of $r(\cdot)$.

There are many possible formulations of the above fitting problem. We describe one: We shall use weighted least-squares as a distance measure, thus the "best-fitting" $\underset{\sim}{r}$ minimizes

$$D(\hat{\underset{\sim}{r}}, \underset{\sim}{r}) = \sum_{i=1}^{k} w_i (\hat{r}_i - r_i)^2 \tag{2.6}$$

subject to a complete monotonicity constraint. Some preliminary numerical work indicates that the very high order differences in the constraints (2.5) have little effect on the solution; since these contribute to the ill-conditioned nature of the problem we truncate the constraint set at a maximum difference d (typically 4, 5, or 6). Furthermore, many of the constraints are redundant, e.g., $\Delta r_i \leq 0$ and $\Delta^2 r_{i+1} \geq 0$ implies that $\Delta r_{i+1} \geq 0$. Finally it is unnecessary to constrain the sequence infinitely far into the future; we shall require constraints on $\underset{\sim}{r}$ only ℓ points into the future.

The constraints are thus

$$(2.7)\qquad \begin{aligned} &(-1)^d \Delta^d r_i \geq 0, \quad && d+1 \leq i \leq k+\ell \\ &(-1)^n \Delta^n r_{k+\ell} \geq 0, \quad && 0 \leq n \leq d-1. \end{aligned}$$

So we have a linearly-constrained quadratic programming problem: Minimize (2.6) subject to (2.7). Nagel et al. (1984) have considered this problem with $d = 1$, the isotone regression case. McWilliams (1979) has also investigated the isotone case.

The quadratic programming algorithm presented in this paper requires that all components of the solution vector appear in the objective function. Thus if we wish to constrain the solution into the future ($\ell > 0$ in equation 2.7) it is necessary to work with equivalent modified sets of constraints. We find constraints on $\{r_1, r_2, \cdots, r_k\}$ such that any solution satisfying these constraints has a feasible extension to $\{r_{k+1}, \cdots, r_{k+\ell}\}$ which satisfies (2.7). For the cases $d = 1$ and $d = 2$, the constraints on $\{r_1, r_2, \cdots, r_k\}$ are simply those in (2.7) with $\ell = 0$ and any feasible solution can then be extended into the future by letting $r_{k+i} = r_k$, $i = 1, 2, \cdots, \ell$; thus in these cases the future constraints have no effect. However, for $d \geq 3$, there are solutions which satisfy (2.7) with $\ell = 0$ but not with $\ell > 0$. The following propositions give the augmented constraints on $\{r_1, \cdots, r_k\}$ needed for the cases $d = 3$ and $d = 4$. The form of these constraints for $d \geq 5$ is unknown; but the following propositions can be used as a partial solution (necessary conditions but not sufficient).

PROPOSITION 2.1. *Consider the constraints (2.7) with* $d = 3$ *and* *any fixed* $\ell > 0$. *A solution* $(r_1, r_2, \cdots, r_k)$ *which satisfies* (2.7) *with* $\ell = 0$ *can be extended to a vector* $(r_1, r_2, \cdots, r_k, r_{k+1}, \cdots, r_{k+\ell})$ *which satisfies* (2.7) *with* $\ell > 0$ *if and only if*

$$\frac{j(j+1)}{2} \Delta^2 r_k + j\Delta r_k + r_k \geq 0, \tag{2.8}$$

$j = 1, 2, \cdots, \ell$.

PROOF. Proof of necessity is by induction. Let $\ell = 1$. From the definition of the difference operator we get

$$\Delta^3 r_{k+1} = r_{k+1} - r_k - \Delta r_k - \Delta^2 r_k. \tag{2.9}$$

A feasible extension must have $\Delta^3 r_{k+1} \leq 0$ and $r_{k+1} \geq 0$; therefore (2.9) implies

$$r_k + \Delta r_k + \Delta^2 r_k \geq 0 \tag{2.10}$$

must hold. Now assume (2.8) is necessary for extending $\ell-1$ points into the future; in particular, extension of $(r_1, r_2, \cdots, r_{k+1})$ to $(r_1, r_2, \cdots, r_{k+1}, \cdots, r_{k+\ell})$. Thus, by inductive hypothesis,

$$\frac{j(j+1)}{2} \Delta^2 r_{k+1} = j\Delta r_{k+1} + r_{k+1} \geq 0, \tag{2.11}$$

$j = 1, 2, \cdots, \ell-1$, is necessary. We now consider extending $(r_1, r_2, \cdots, r_k)$ to $(r_1, r_2, \cdots, r_k, \cdots, r_{k+\ell})$. Since $\Delta^3 r_{k+1} \leq 0$, we get

$$\Delta^2 r_{k+1} \leq \Delta^2 r_k \tag{2.12}$$

and from (2.9) we get

$$r_{k+1} \le \Delta^2 r_k + \Delta r_k + r_k. \tag{2.13}$$

And by definition of Δr_{k+1}, (2.13) becomes

$$\Delta r_{k+1} \le \Delta^2 r_k + \Delta r_k. \tag{2.14}$$

Substituting (2.12), (2.13) and (2.14) into (2.11) gives

$$\frac{(j+1)(j+2)}{2} \Delta^2 r_k + (j+1)\Delta r_k + r_k \ge 0, \tag{2.15}$$

$j = 1,2,\cdots,\ell-1$, whenever a feasible extension exists. From (2.13) we see that (2.15) also holds for $j = 0$, which gives (2.8).

Sufficiency of (2.8) is demonstrated by constructing a feasible extension. Let

$$y_{k+j} = r_k + j\Delta r_k + \frac{j(j+1)}{2} \Delta^2 r_k, \quad j = 1,2,\cdots,\ell \tag{2.16}$$

and define m as

$$m = \max(j : \Delta y_{k+j} \le 0). \tag{2.17}$$

Then a feasible extension is

$$r_{k+j} = \begin{cases} y_{k+j}, & j = 1,2,\cdots,m \\ y_{k+m}, & j = m+1,\cdots,\ell. \end{cases} \tag{2.18}$$

This is clearly feasible: From (2.8), $y_{k+j} \geq 0$, therefore $r_{k+j} \geq 0$. From (2.17), it follows that $\Delta r_{k+j} \leq 0$ for $j = 1,2,\cdots,m$, and $\Delta r_{k+j} = 0$ for $j = m+1,\cdots,\ell$. Clearly, y_{k+j}, $j = 1,2,\cdots,\ell$ is a quadratic with $\Delta^2 y_{k+j} = \Delta^2 r_k \geq 0$, so $\Delta^2 r_{k+j} = \Delta^2 r_k$, $j = 1,2,\cdots,m$; $\Delta^2 r_{k+m+1} = \Delta r_{k+m+1} - \Delta r_{k+m} = -\Delta r_{k+m} \geq 0$ (by (2.17) and (2.18)); $\Delta^2 r_{k+j} = 0$, $j = m+2,\cdots,\ell$. Finally, $\Delta^3 r_{k+1} = 0$, $j = 1,2,\cdots,m$; $\Delta^3 r_{k+m+1} = \Delta^2 r_{k+m+1} - \Delta^2 r_{k+m} = -\Delta r_{k+m} - \Delta^2 r_{k+m} \leq 0$; $\Delta^3 r_{k+m+2} = \Delta^2 r_{k+m+2} - \Delta^2 r_{k+m+1} = -\Delta^2 r_{k+m+1} \leq 0$; and $\Delta^3 r_{k+j} = 0$, $j = m+3,\cdots,\ell$. This completes the proof of Proposition 2.1.

PROPOSITION 2.2. *Consider the constraints* (2.7) *with* $d = 4$ *and any fixed* $\ell > 0$. *A solution* $(r_1, r_2, \cdots, r_k)$ *which satisfies* (2.7) *with* $\ell = 0$ *can be extended to a vector* $(r_1, r_2, \cdots, r_k, r_{k+1}, \cdots, r_{k+\ell})$ *which satisfies* (2.7) *with* $\ell > 0$ *if and only if*

$$\frac{j(j+1)}{2}\Delta^3 r_k + j\Delta^2 r_k + \Delta r_k \leq 0, \quad j = 1,2,\cdots,\ell, \tag{2.19}$$

$$\frac{j(j-1)}{6}\Delta^2 r_k + \frac{2(j-1)}{3}\Delta r_k + r_k \geq 0, \quad j = 2,3,\cdots,\ell, \tag{2.20}$$

$$\frac{\ell(\ell+1)}{2}\Delta^2 r_k + k\Delta r_k + r_k \geq 0. \tag{2.21}$$

PROOF. Conditions (2.19) and (2.21) are necessary from Proposition 2.1. To verify (2.20), consider the extension j points into the future; the following inequalities are necessary:

$$
\begin{aligned}
0 &\le \Delta^4 r_{k+1} = r_{k+1}-r_k-\Delta r_k-\Delta^2 r_k-\Delta^3 r_k \\
0 &\le \Delta^4 r_{k+2} = r_{k+2}-4r_{k+1}+3r_k+2\Delta r_k+\Delta^2 r_k \\
0 &\le \Delta^4 r_{k+3} = r_{k+3}-4r_{k+2}+6r_{k+1}-3r_k-\Delta r_k \\
0 &\le \Delta^4 r_{k+i} = r_{k+i}-4r_{k+i-1}+6r_{k+i-2}-4r_{k+i-3}+r_{k+i-4}, \quad 4 \le i \le j \\
0 &\le -\Delta^3 r_{k+j} = -(r_{k+j}-3r_{k+j-1}+3r_{k+j-2}-r_{k+j-3}) \\
0 &\le \Delta^2 r_{k+j} = r_{k+j}-2r_{k+j-1}+r_{k+j-2} \\
0 &\le -\Delta r_{k+j} = -(r_{k+j}-r_{k+j-1}) \\
0 &\le r_{k+j}.
\end{aligned} \tag{2.22}
$$

Define

$$
\beta_i = \begin{cases} (j-i+1)(j-i+2)(i-1)/6, & i = 1,2,\cdots,j+3 \\ 1 & i = j+4 \end{cases} \tag{2.23}
$$

and consider, from (2.22),

$$
0 \le \sum_{i=1}^{j} \beta_i \Delta^4 r_{k+i} - \beta_{j+1}\Delta^3 r_{k+j} + \beta_{j+2}\Delta^2 r_{k+j} - \beta_{j+3}\Delta r_{k+j} + \beta_{j+4} r_{k+j}. \tag{2.24}
$$

From (2.23)

$$
\begin{aligned}
&\beta_i - 4\beta_{i+1} + 6\beta_{i+2} - 4\beta_{i+3} + \beta_{i+4} = 0, \quad i < j \\
&\beta_{j+1} = 0 \\
&\beta_{j+2} = 0.
\end{aligned}
$$

So (2.24) becomes

$$(2.25) \qquad 0 \le \beta_2(3r_k+2\Delta r_k+\Delta^2 r_k) + \beta_3(-3r_k-\Delta r_k) + \beta_4 r_k,$$

which is exactly the same as (2.20). (Note that strictly speaking the above argument holds for $j \ge 4$; similar simpler arguments hold for $j = 2,3.$).

The sufficiency of the above conditions is demonstrated by giving feasible extensions of $(r_1,\cdots,r_k)$ to $(r_1,\cdots,r_{k+\ell})$. There are two cases: in the first, the extension is a quadratic; in the second it is a cubic which becomes constant when its slope hits zero. If

$$(2.26) \qquad \ell\Delta^2 r_k + \Delta r_k \le 0,$$

define

$$(2.27) \qquad r_{k+j} = \frac{j(j+1)}{2}\,\Delta^2 r_k + j\Delta r_k + r_k, \qquad j = 1,2,\cdots,\ell.$$

Then

$$\Delta r_{k+j} = j\Delta^2 r_k + \Delta r_k \le 0, \qquad j = 1,2,\cdots,\ell$$

from (2.26); and it suffices to check $r_{k+j} \ge 0$ only when $j = \ell$, which follows from (2.21). Higher-order differences are shown to be feasible using definitions and routine algebra.

If (2.26) does not hold, let

$$(2.28) \qquad m = \min(\ell,[-2\Delta r_k/\Delta^2 r_k] + 1),$$

where $[\cdot]$ is the greatest integer function. On the interval $j = 1,\cdots,m-1$, define r_{k+j} as a cubic and then as a constant for $j = m-1,\cdots,\ell$. The third difference is

$$(2.29)\qquad \Delta^3 r_{k+j} = -\frac{2\Delta r_k}{m(m+1)} - \frac{2\Delta^2 r_k}{m+1}, \qquad j = 1,2,\cdots,m.$$

Summing gives

$$(2.30)\qquad \begin{aligned} \Delta^2 r_{k+j} &= \Delta^2 r_k + \sum_{i=1}^{j} \Delta^3 r_{k+i} \\ &= -\frac{2j}{m(m+1)} \Delta r_k + \left[1-\frac{2j}{m+1}\right]\Delta^2 r_k, \qquad j = 1,2,\cdots,m, \end{aligned}$$

$$(2.31)\qquad \begin{aligned} \Delta r_{k+j} &= \Delta r_k + \sum_{i=1}^{j} \Delta^2 r_{k+i} \\ &= \left[1 - \frac{j(j+1)}{m(m+1)}\right]\Delta r_k + \frac{j(m-j)}{m+1} \Delta^2 r_k, \qquad j = 1,2,\cdots,m, \end{aligned}$$

$$(2.32)\qquad \Delta r_{k+j} = 0, \qquad j = m+1, m+2, \cdots, \ell.$$

Finally

$$(2.33)\qquad r_{k+j} = \begin{cases} r_k + \left[j - \dfrac{j(j+1)(j+2)}{3m(m+1)}\right] \Delta r_k + \dfrac{j(j+1)(3m-2j-1)}{6(m+1)} \Delta^2 r_k, & j = 1,2,\cdots,m-1 \\ r_{k+m-1}, & j = m, m+1, \cdots, \ell. \end{cases}$$

The third difference is negative by definition of m; thus the second difference need only be checked for $j = m$ which is positive by definition of m (m was defined to meet this condition); thus the first difference is an increasing function for $j = 1,2,\cdots,m$, and also since it equals zero for $j = m$ the constraint is satisfied. Furthermore, the differences for $j = m+1,\cdots,\ell$ can be shown to have the correct signs by algebraic manipulation. So it suffices to verify that

$$(2.34) \qquad 0 \le r_{k+m-1} = r_k + \frac{2(m-1)}{3} \Delta r_k + \frac{m(m-1)}{6} \Delta^2 r_k,$$

which is true from condition (2.20). Finally,

$$(2.35) \qquad \Delta^4 r_{k+1} = \Delta^3 r_{k+1} - \Delta^3 r_k = -\frac{2\Delta r_k}{m(m+1)} - \frac{2\Delta^2 r_k}{m+1} - \Delta^3 r_k,$$

which is nonnegative by condition (2.19). This completes the proof of Proposition 2.2.

3. OUTLINE OF AN ACTIVE CONSTRAINT ALGORITHM. The problem of concern is a particular case of the quadratic programming problem

$$(3.1) \qquad \text{(QP)} \quad \underset{x \in R^n}{\text{minimize}} \quad f(x) = 1/2\, x^T G x + c^T x$$

$$(3.2) \qquad \text{subject to} \quad Ax \ge b$$

where x and c are n-vectors, G is an $n \times n$ symmetric matrix and A is an $m \times n$ matrix. Although the matrix G is a diagonal matrix for the least squares regression problem, we shall assume throughout the following that G is any symmetric positive definite matrix. The gradient of $f(x)$, denoted by $\nabla f(x)$ is computed by

$$\nabla f(x) = Gx + c.$$

Let a_j be the $j^{\underline{th}}$ row of A. The constraint $a_j x \le b_j$ is said to be active at the point $\hat{x}$ if it is satisfied exactly, that is, $a_j\hat{x} = b_j$. A point $\hat{x}$ is feasible to the problem if it satisfies all problem constraints, i.e., $A\hat{x} \geq b$.

Let $\hat{x}$ be a point in R^n. Let $\hat{A}$ $(r \times n)$ (where $r \leq m$) be the matrix of constraints active at $\hat{x}$, and let $\hat{b}$ be the corresponding right hand side vector. Assume also that $\hat{A}$ is of full row rank. Then the necessary and sufficient conditions for $\hat{x}$ to be the (unique) solution to (QP) are:

(3.3) (C1) $A\hat{x} \geq b$ (feasibility)

(3.4) (C2) There exists an r-vector λ such that $\hat{A}^T\lambda = \nabla f(\hat{x})$ (λ is defined as the vector of Lagrange multipliers)

(3.5) (C3) $\lambda \geq 0$.

The underlying concept behind active set methods is the following: Suppose that the correct active set (i.e., the set of constraints that are active at the optimum) is known. Then the solution to (QP) is also the solution to the equality constrained quadratic problem

$$\text{(EQP)} \quad \text{minimize } f(x) = 1/2\ x^TGx + c^Tx \qquad (3.6)$$
$$\text{subject to} \quad \hat{A}x = \hat{b}.$$

Active set methods generate a sequence of problems of the form (EQP). In each such subproblem, the quadratic objective is to be minimized over the manifold obtained by fixing a subset of the constraints (known as the working set) as equalities. In general, the working set includes only constraints that are exactly satisfied at the current point, and subproblems differ only in the constraints that are added or deleted from the working set.

An important feature of active set methods is that they are "primal feasible"; that is, they start with a feasible point and generate a sequence of feasible points. The optimization of each subproblem is performed by consecutively minimizing the objective along feasible directions — directions which maintain feasibility of the equality constraints. The direction of movement at each point should be a direction of descent, i.e., with a negative directional derivative, which insures that the objective will initially decrease; if no such direction is available, then the point is optimal for the current subproblem. The line minimization itself can terminate in one of two ways: either a local minimum of the objective along the line is obtained at a point feasible for (QP), or the minimization is terminated due to the encounter of a constraint which is not in the working set (further movement along this line would cause that particular constraint to be violated). The most commonly used approach in the latter case, is to add the newly encountered constraint to the working set, and repeat the process of solving (EQP) starting from the new point and using the new working set.

Once the solution to an equality constrained quadratic subproblem is obtained, the associated vector of multipliers is computed, using the over-determined system of equations (3.4). If this vector is nonnegative, then the optimality conditions are satisfied and the current point is the optimum for (QP). If, however, some multiplier is negative, then the current working set is not the correct active set, and the objective can be improved by dropping the constraint corresponding to this negative multiplier from the working set. The entire process is now repeated with the new working space.

We now discuss a method for obtaining the direction of movement.

Let $\hat{x}$ be the current point, $\hat{A}$ the constraint matrix associated with the current working set, and $\hat{b}$ the corresponding right-hand side vector (hence $\hat{A}\hat{x} = \hat{b}$). Our aim is to find a direction of movement p such that the vector $\hat{x}+p$ solves (EQP). If $\hat{x}+p$ is to satisfy $\hat{A}(\hat{x}+p) = \hat{b}$, then the direction p should satisfy

$$(3.7) \qquad \hat{A}p = 0.$$

Thus, the direction p lies in a subspace defined by the working set. Assume that the $r \times n$ matrix $\hat{A}$ has full row rank. Let Z be any $n \times t$ matrix $(t \geq n-r)$ whose columns span the null space of $\hat{A}$. Then equation (3.7) is equivalent to

$$(3.8) \qquad p = Zv.$$

for some t-vector v. Using (3.8), and substituting $\hat{x}+p$ into (3.6), we get the equivalent *unconstrained* optimization problem

$$(3.9) \qquad \underset{v \in R^t}{\text{minimize}}\ F(v) = 1/2\ v^T(Z^TGZ)v + \nabla f(\hat{x})^T Zv.$$

The solution to (3.9) is obtained by setting the gradient of F(v) to zero. It is the point v which satisfies the system of equations

$$(3.10) \qquad (Z^TGZ)v = -Z^T\nabla f(\hat{x}).$$

Once v is obtained, the vector p is computed through (3.8).

The matrix Z^TGZ of equation (3.10) is known as the projected

Hessian matrix, while the vector $Z^T\nabla f(\hat{x})$ is the projected gradient. Methods for which the direction of movement is determined by (3.10) are known as Newton-type methods.

Various numerical linear algebra techniques exist for the solution of system (3.10). Here we consider two approaches, which differ only in the method for representing the matrix Z. As we shall see, the first implementation is by far superior to the second.

The Orthogonal Factorization Approach. Under this approach, the matrix $\hat{A}$ is factorized into the form

$$\hat{A} = [L,0]Q$$

where L is an $r\times r$ lower triangular matrix and Q is an orthogonal matrix. If the matrix Q is partitioned into

$$Q = \begin{bmatrix} Q_1 \\ Q_2 \end{bmatrix}$$

where Q_1 is $r\times n$ and Q_2 is $(n-r)\times n$, then since $\hat{A}Q_2^T = 0$, the matrix Z can be taken as $Z = Q_2^T$. Given this matrix Z, the most efficient way to solve (3.10) is to compute the Cholesky factorization of the projected Hessian

$$Z^TGZ = U^TU$$

where U is an upper triangular matrix. Updating schemes are used from one iteration to another, so that the factors do not have to be recomputed from scratch at each iteration. For further detail, see Gill and Murray (1977).

The major advantage of this approach is its numerical stability. Since the ℓ_2 condition number of a matrix is unchanged when that matrix is either pre- or post-multiplied by an orthogonal matrix, the condition number of Z^TGZ cannot be worse than that of G. Hence, if G is well conditioned, the solution process is very stable numerically.

The Projection Matrix Approach. Under this approach we choose $Z = I-\hat{A}^T(\hat{A}\hat{A}^T)^{-1}\hat{A}$. By substituting the latter into (3.10) it can be shown that the direction of search p is given by

$$p = G^{-1}(\hat{A}^T\lambda-\nabla f(\hat{x})) \tag{3.11}$$

where λ is the solution to the system of equations

$$(\hat{A}G^{-1}\hat{A}^T)\lambda = \hat{A}G^{-1}\nabla f(\hat{x}). \tag{3.12}$$

As in the previous case, a Cholesky factorization is used for the matrix $\hat{A}G^{-1}\hat{A}^T$, in order to solve this system.

Although the amount of computational work involved in this approach is significantly smaller than that of the orthogonal factorization approach, the method has an underlying computational weakness due to the ill conditioning of the matrix $\hat{A}G^{-1}\hat{A}^T$. Thus, for example, if G is the indentity matrix, the aforementioned matrix has condition number of order $(k(\hat{A}))^2$. This could lead to instability in the solution of (3.12) which in turn could lead to an unsatisfactory vector p in (3.11). Thus, due to round off error the vector p will no longer satisfy $\hat{A}p = 0$, and the new point obtained could be infeasible.

4. PERFORMANCE OF ALGORITHM. We now discuss how different approaches to solving our original problem (*) perform. We shall not discuss the statistical merits of using this kind of regression model for making inferences. The approach appears promising (see Miller and Sofer (1985)); however, there are many refinements in the statistical formulation which must be considered: How many discrete intervals should be used? What weights should be used in the objective function? Should all the failure data be used or just the recent data? And so forth.

We now present some indication of how the algorithms perform by applying them to 20 different sets of failure data. The data was Monte Carlo generated from 3 members of a family of nonhomogeneous Poisson processes with cumulative mean functions.

$$M(t) = 40\,\frac{\log(\beta t+1)}{\log(100\beta+1)}, \quad 0 \le t \le 100, \tag{4.1}$$

$\beta = .0429,\ .461,\ 31.0,$ and from a homogeneous Poisson process with cumulative mean function

$$M(t) = 40t/100, \quad 0 \le t \le 100, \tag{4.2}$$

which actually corresponds to the limiting case of $\beta = 0$ in (4.1). Data was collected over the interval $[0,100]$ so the expected number of failures is 40 for each model. Five independent replicates were generated from each of the above models for a total of 20 data sets. We then considered three different discretizations of the data, i.e., (2.2): 10, 20, or 30 intervals. The rational for using such models to generate data is that they are the Musa-Okumoto (1984) models, which are often reasonable models for software reliability growth; see

Miller (1986) and Miller and Sofer (1985) for further discussion.

So we have 20 sets of data for which we want to solve the 10, 20, or 30 dimensional version of our optimization problem (*) with nonredundant constraints (1.2) for $d = 2$, 3, or 4. We do not consider future constraints here and we use unit weights in the objective function. We would like a fast, accurate solution method for this type of problem.

The first method we considered is the active-set approach described in Section 3. We first used the projection matrix implementation described there. We found this to be unacceptable. The numerical instability led to "solutions" which were slightly infeasible in the sense that the constraints (1.2) were violated by a very small amount, say 10^{-7}. This slight violation tended to reflect itself in a gross violation of some of the redundant constraints in (1.1) which are not included in (1.2). We believe that this behavior is related to the ill-conditioned nature of the problem. In Table 4.1 we show such a solution: the second column is the data $(r_1, r_2, \cdots, r_{30})$ and the third column is the solution $(x_1, x_2, \cdots, x_{30})$; the maximum constraint violation is 10^{-7} in this case, but note the violation of monotonicity. We draw two conclusions from this example: First, the active-set approach needs a more numerically stable algebraic implementation to prevent infeasibility. Secondly, if infeasible solutions are allowed as approximate solutions, constraint violations must be very, very small. This approach was used by Miller and Sofer (1985).

Table 4.1. An Infeasible Approximate Solution*

Interval Number	Rate	Solution
1	1.9788	2.1244
2	1.6974	1.2202
3	.5515	.7850
4	.2052	.6062
5	.2405	.5517
6	.5996	.5464
7	.5796	.5536
8	.5896	.5590
9	.7389	.5599
10	.9011	.5572
11	.7586	.5517
12	.2743	.5439
13	.2745	.5343
14	.5321	.5235
15	.7688	.5117
16	.9375	.4991
⋮	⋮	⋮

*Calculated using second implementation of Active-set approach. Maximum constraint violation is 10^{-7}. $k = 40$, $d = 7$.

We then considered the orthogonal matrix implementation of the active-set approach in Section 3. This approach is slightly slower, conceptually more difficult and involves more complicated code; however, it performs much better. It seems to be an acceptable method for the problem and application addressed in this paper. Some performance statistics are shown in Table 4.2 for the 20 data sets mentioned above. These numbers reflect satisfactory performance for our purposes so we felt it unnecessary to estimate the performance characteristics more accurately.

Table 4.2. Performance of active-set algorithm on 20 data sets.

Number of Intervals	Maximum Difference	Average Iterates	Average Time* per Problem	Average Time* per Iterate
10	2	13.0	.036	.0028
10	3	13.4	.040	.0030
10	4	14.9	.041	.0027
20	2	15.7	.097	.0061
20	3	18.1	.110	.0060
20	4	21.0	.128	.0061
30	2	18.1	.199	.0110
30	3	23.2	.240	.0103
30	4	28.6	.284	.0099

*Times are given in seconds. Execution was performed with an IBM 4341 CPU with approximately 1.2 mips.

Finally, we thought it would be interesting to see how at least one other algorithm would handle this problem. We chose Dykstra's (1983) algorithm because it is easy to program and because several people have reported favorable experiences with it to us in private communications. We chose a stopping rule that said to stop when the maximum constraint violation in (1.2) becomes less than 10^{-7}; we chose such a small value because of the experience we had in Table 4.1. We found that many iterations were required to meet this stopping rule; so we decided to stop after 10,000 iterations regardless. We applied the algorithm to our 20 data sets; there were 9 cases corresponding to 10, 20, or 30 intervals and $d = 2, 3$, or 4, the highest order difference. The only case in which all 20 cases converged in fewer than 10,000 iterations was $d = 2$ with 10 intervals; the average number of iterations was 1294, but the distribution was highly skewed with 15 cases less than 1000 and 7 cases less than 100 iterations; the average time per problem was 2.13 CPU seconds, for an average of .0016 seconds per iteration. In the remaining 8 cases, 27.5% of the problems converged in less than 10,000 iterations. In the analyses based on 20 and 30 intervals, 60% of the $d = 3$ cases converged and only 1 of the $d = 4$ cases converged. In the 30 interval, $d = 4$ case it required an average of 94 CPU seconds to do the 10,000 iterations on each problem. The stopping criterion using 10^{-7} may seem extreme; however, in the cases which converged we observed a geometric sequence so that a criterion using $10^{-3.5}$ would require approximately one-half the execution time. We did observe, however, cases which achieved this lesser criterion in 10,000 iterations but were clearly unacceptable. Table 4.3 shows such a solution. Finally, Dykstra's algorithm appears to be quite data

Table 4.3. Data and solutions using Dykstra's algorithm (10,000 iterations, .00093 maximum constraint violation).

Interval Number	Rate	Solution
1	0.261197	0.435425
2	0.710503	0.411107
3	0.310693	0.386801
4	0.240826	0.362504
5	0.194263	0.370655
6	0.181977	0.378753
7	0.273743	0.386741
8	0.543812	0.394554
9	0.285000	0.402116
10	0.180889	0.409346
11	0.326807	0.416160
12	0.326649	0.422476
13	0.191235	0.428210
14	0.714396	0.433287
15	1.133522	0.437637
16	0.994115	0.441197
17	0.222963	0.443916
18	0.100874	0.445754
19	0.823767	0.446681
20	0.282771	0.446681

*Data is generated from a homogeneous Poisson process.

dependent; it performed much better for the cases of the Musa-Okumoto model with $\beta > 0$ than it did for data from the homogeneous Poisson process.

5. **CONCLUSIONS.** The paper presents the numerical difficulties associated with the problem of least squares regression under higher order difference constraints. As the problem size grows, it becomes increasingly ill conditioned. Thus, a numerically stable algorithm should be used for solving the problem. In our experience, the active set method using orthogonal matrix factorizations is a viable algorithm for solving this type of problem.

APPENDIX A

The Condition Number of the Matrix A. Let $\|\cdot\|_\alpha$ denote any matrix norm. Let A be a nonsingular matrix. The condition number of A with respect to this norm is defined by

$$\kappa_\alpha(A) = \|A\|_\alpha \cdot \|A^{-1}\|_\alpha. \tag{A.1}$$

Several results exist which relate the condition number of A to the sensitivity of the linear system $Ax = b$. For example, suppose that the right hand side of this system is perturbed by Δb, and let $x+\Delta x$ be the solution of the perturbed system

$$A(x+\Delta x) = b+\Delta b.$$

Then using any vector norm and consistent matrix norm it can be shown (see Golub and Van Loan (1983)) that

$$\frac{\|\Delta x\|}{\|x\|} \le \kappa(A) \frac{\|\Delta b\|}{\|b\|},$$

and thus in effect, $\kappa(A)$ determines the sensitivity of the system.

Matrices with small condition numbers (close to 1) are said to be well conditioned, while matrices with large condition numbers are said to be ill conditioned. Although this is a norm dependent statement, it can be shown (see Golub and Van Loan (1983)) that any two condition numbers $\kappa_\alpha(\cdot)$ and $\kappa_\beta(\cdot)$ are equivalent, in the sense that there exist positive constants c_1 and c_2 for which

$$c_1\kappa_\alpha(A) \le \kappa_\beta(A) \le c_2\kappa_\alpha(A).$$

In order to determine the condition number of the matrix A defined by (1.3) we first prove the following proposition.

PROPOSITION A.1. *Consider the matrix* A *defined by* (1.3). *Let* B *be the matrix defined by*

$$\text{(A.2)} \qquad b_{ij} = \begin{cases} \binom{d-1+j-i}{j-i} & \begin{matrix} 1 \le i \le n-d \\ i \le j \le n-d \end{matrix} \\ \binom{n-i}{j-i} & \begin{matrix} i \le i \le n-d \\ n-d+1 \le j \le n \end{matrix} \\ \binom{n-i}{j-i} & \begin{matrix} n-d+1 \le i \le n \\ i \le j \le n-d \end{matrix} \\ 0 & \textit{otherwise}. \end{cases}$$

Then $B = A^{-1}$.

PROOF: Let $C = AB$. Since both A and B are upper triangular with unit diagonal, C is also upper triangular and $c_{ii} = 1$ for

$i = 1,\cdots,n$. Thus it remains to prove that $c_{ij} = 0$ for $i < j$.

Case (a): $1 \le i \le n-d$, $i < j \le i+d$.

Then

$$c_{ij} = \sum_{k=1}^{n} a_{ik}b_{kj} = \sum_{k=i}^{i+d} (-1)^{k-i} \binom{d}{k-i} \binom{d-1+j-k}{j-k}$$

$$= \sum_{k=0}^{d} (-1)^{k} \binom{d}{k} \binom{d+(j-i-k)-1}{j-i-k}.$$

We now invoke the following identity (see Feller (1968))

(A.3) $$\binom{-a}{m} = (-1)^{m} \binom{a+m-1}{m}$$

to get

$$c_{ij} = \sum_{k=0}^{d} (-1)^{j-i} \binom{d}{k} \binom{-d}{j-i-k}.$$

By comparing the zero coefficients of t^{j-i} $(j > i)$ in $(1-t)^{d}(1-t)^{-d} = 1$, we conclude that $c_{ij} = 0$.

Case (b): $1 \le i \le n-d$, $n-d+1 \le j \le n$.

$$c_{ij} = \sum_{k=i}^{\min(j,i+d)} (-1)^{k-i} \binom{d}{k-i} \binom{n-k}{j-k}$$

$$= \sum_{k=0}^{\min(j-i,d)} (-1)^{k} \binom{d}{k} \binom{n-i-k}{j-i-k}.$$

We now use (A.3) to get

$$c_{ij} = \sum_{k=0}^{\min(j-i,d)} (-1)^{j-i} \begin{bmatrix} d \\ k \end{bmatrix} \begin{bmatrix} -(n+1-j) \\ j-i-k \end{bmatrix}.$$

By comparing the zero coefficient of t^{j-i} $(j > i)$ in $(1-t)^d(1-t)^{-(n+1-j)} = (1-t)^{j-(n-d)-1}$ (recall that $i \le n-d$), we conclude that $c_{ij} = 0$.

Case (c): $n-d+1 \le i \le n, \quad i \le j \le n-d.$

We use (A.3) again to obtain

$$\begin{aligned} c_{ij} &= \sum_{k=i}^{n} (-1)^{k-i} \begin{bmatrix} n-i \\ k-i \end{bmatrix} \begin{bmatrix} n-i \\ j-i \end{bmatrix} \\ &= \sum_{k=0}^{n-i} (-1)^{k} \begin{bmatrix} n-i \\ k \end{bmatrix} \begin{bmatrix} n-i-k \\ j-i-k \end{bmatrix} \\ &= \sum_{k=0}^{n-i} (-1)^{j-i} \begin{bmatrix} n-i \\ k \end{bmatrix} \begin{bmatrix} -(n+1-j) \\ j-i-k \end{bmatrix}. \end{aligned}$$

Comparing the zero coefficient of t^{j-i} in the identity $(1-t)^{n-i}(1-t)^{-(n+1-j)} = (1-t)^{j-i-1}$, we conclude that $c_{ij} = 0$.

In order to compute the condition number of A we shall use the $\|\cdot\|_\infty$ norm which is defined by

$$\|A\|_\infty = \max_i \left\{ \sum_{j=1}^{n} |a_{ij}| \right\}.$$

PROPOSITION A.2. *Let* A *be the matrix defined by* (1.3). *Then* $\kappa_\infty(A)$ *is given by*

(A.4)
$$\kappa_\infty(A) = 2^d \cdot \left[\binom{n-1}{d} + \sum_{j=0}^{n-d} \binom{n-1}{j}\right].$$

PROOF. Since

$$\sum_{j=1}^{n} |a_{ij}| = 2^d \quad \text{for} \quad i \le i \le n-d$$

and

$$\sum_{j=1}^{n} |a_{ij}| = 2^{n-i} \quad \text{for} \quad n-d+1 \le i \le n,$$

we get that

(A.5)
$$\|A\|_\infty = 2^d.$$

In order to compute $\|A^{-1}\|_\infty$ we note that the first row of the matrix A^{-1} dominates all other rows. Thus

$$\|A^{-1}\|_\infty = \sum_{j=1}^{n-d} \binom{d-2+j}{j-1} + \sum_{j=n-d+1}^{n} \binom{n-1}{j-1}.$$

We now use the identity

$$\sum_{j=0}^{r} \binom{j+k-1}{j} = \binom{r+k}{k}$$

(see Feller (1968)) with $r = n-d-1$, $k = d$ to obtain

(A.6)
$$\|A^{-1}\|_\infty = \binom{n-1}{d} + \sum_{j=0}^{n-d} \binom{n-1}{j}.$$

Thus, substituting (A.5) and (A.6) into (A.1) we get the desired result.

By examining eq. (A.4) it is not difficult to see that the matrix becomes increasingly ill conditioned as n and d increase. Even for small sized problems this matrix is extremely ill conditioned. For example, $n = 40$ and $d = 2$ will give the condition number $\kappa_\infty(A) = 3124$, while $n = 40$ and $d = 3$ yield $\kappa_\infty(A) = 79360$.

REFERENCES

Barlow, R.E., Bartholomew, D.J., Bremner, J.M. and Brunk, H.D. (1972). *Statistical Inference Under Order Restrictions*. Wiley, New York.

Crow, L.H. (1974). Reliability analyses for complex, reparable systems. *Reliability and Biometry* (F. Proschan and R.J. Serfling, eds.), SIAM, Philadelphia, 379-410.

Duane, J.T. (1964). Learning curve approach to reliability monitoring. *IEEE Trans. Aerospace* **2**, 563-566.

Dykstra, R.L. (1983). An algorithm for restricted least squares regression. *J. Amer. Statist. Assoc.* **78**, no. 384, 837-842.

Feller, W. (1968). *An Introduction to Probability Theory and Its Applications*. Vol. I, 3rd ed. Wiley, New York.

Feller, W. (1971). *An Introduction to Probability Theory and Its Applications*. Vol. II, 2nd ed. Wiley, New York.

Gill, P.E. and Murray, W. (1977). Modifications of matrix factorization after a rank-one change, in *State of the Art in Numerical Analysis* (D. Jacobs, ed.). Academic Press.

Gill, P.E. and Murray, W. (1978). Numerically stable methods for quadratic programming. *Math. Programming*, **14**, 349-372.

Goel, A.K. and Okumoto, K. (1979). Time-dependent error detection rate model for software reliability and other performance measures. *IEEE Trans. Rel.* **R-28**, 206-211.

Golub, G.H. and Van Loan, C.F. (1983). *Matrix Computations*. The Johns Hopkins University Press, Baltimore, Maryland.

Jelinsky, Z. and Moranda, P. (1972). Software reliability research. *Statistical Computer Performance Evaluation* (W. Ferberger, ed.). Academic Press, New York, 465-484.

Littlewood, B. (1981). Software reliability growth: A model for fault-removal in computer-programs and hardware-design. *IEEE*

Trans. Rel. **R-30**, 313-320.

McCormick, G.P. (1983). *Nonlinear Programming: Theory, Algorithms and Applications*. Wiley, New York.

McWilliams, T.P. (1979). Properties of isotonic estimators of mean lifetime in a simple prototype development model, TR-194. Department of Operations Research and Department of Statistics, Stanford University, Stanford, California.

Miller, D.R. (1986). Exponential order statistics models for software reliability growth. *IEEE Trans. Software Eng.* **SE-12**, 12-24.

Miller, D.R. and Sofer, A. (1985). Completely monotone regression estimates of software failure rates. *Proceedings of Eighth International Conference on Software Engineering*. IEEE Computer Society Press, Washington, D.C., 343-348.

Musa, J.D. and Okumuto, K. (1984). A logarithmic Poisson execution time model for software reliability measurement. *Proceedings of Seventh International Conference on Software Engineering*. IEEE, New York, 230-238.

Nagel, P.M. Scholtz, F.W. and Skrivan, J.A. (1984). Software reliability: Additional investigations into modeling with replicated experiments. CR-172378, NASA.

Douglas R. Miller
School of Engineering and Applied Science
The George Washington University
Washington, D.C. 20052

Ariela Sofer
System Engineering Department
George Mason University
Fairfax, VA 22030

DOSE-RESPONSE ANALYSIS UNDER UNIMODALITY OF RESPONSE-TO-DOSE[1]

Richard L. Schmoyer
Oak Ridge National Laboratory

AMS 1980 subject classifications: Primary 62P10, secondary 62N05.
Keywords and phrases: accelerated testing, linear interpolation, quantal bioassay.

ABSTRACT

Suppose a dose-response curve π is sigmoidal with mode (inflection point) M. Then the response-to-dose ratio $\pi(x)/x$ is unimodal and maximized at a point $H \geq M$. If $\pi(0)$ is known, if H is known, and if $x_0 \leq x \leq H$, then an upper confidence bound (UCB) for $\pi(x_0)$ can be constructed from a UCB for $\pi(x)$. Unfortunately, H is seldom known in practice. In the setting of quantal bioassay, two lower confidence bounds for H and corresponding UCB's for $\pi(x_0)$ when H is unknown are obtained and contrasted. One is based on a likelihood-ratio statistic; the other on a weighted sum of constraint contrasts. Both tests are deficient in several ways, but have practical utility nevertheless. A maximum-violator algorithm with guaranteed convergence is given for computing the maximum likelihood estimates of the response probabilities subject to unimodality of $\pi(x)/x$.

---

1. This research is sponsored by the Office of Health and Environmental Research, U.S. Department of Energy, under contract DE-AC05-840R21400 with Martin Marietta Energy Systems, Inc.

1. INTRODUCTION. Decisions about the safety of potentially hazardous substances are frequently based on the results of quantal bioassay experiments. In these experiments a substance is administered to animal subjects, which are ostensibly like mine shaft canaries—at least as sensitive to the substance as man. Several animals typically receive the same dose of the substance, and these comprise a group. After a predetermined amount of time, the proportion of subjects in each group that respond in a specified way, for example by dying or by developing a tumor, is recorded. Ordinarily the experiments are accelerated in the sense that the substance being assayed is administered at dose levels that are much higher than the levels of primary interest, which correspond, for example, to levels actually encountered in the environment or workplace. If the experiments were not accelerated, the number of subjects required to elicit one or more responses would usually be prohibitive.

Let π be a dose-response curve, that is, let $\pi(x)$ denote the expected proportion of subjects that respond at a dose $x \geq 0$. The statistical analysis of a quantal bioassay experiment ordinarily involves estimating quantiles of π, for example, the ED 50 (dose x at which $\pi(x) = 1/2$). These quantiles are often compared with quantiles of standard substances, and measures of relative potency are thereby obtained. Unfortunately, it is easy to construct plausible examples in which the test substance appears on the basis of these measures to be much less hazardous than the standard, but with just the opposite occurring at the dose levels of primary interest. For the experiment to reflect the state of knowledge and uncertainty about the test substance at dose levels of interest, it is necessary to have a sensible way to extrapolate, at least qualitatively, from the observed accelerated dose levels to the levels of interest. This implies that some assumption must be made about the lower tail of π.

One such assumption about a dose-response curve is that the curve is a member of a particular parametric family. The problem with this approach, of course, is that a parametric curve that fits the data well over the range of observed doses may be biased absurdly in the region of interest. Furthermore, the appropriateness of a particular parametric model must generally be judged on the basis of its fit in the accelerated region, since theoretical motivations are usually unavailable.

Alternatively, assumptions about the curvature of a dose-response curve can lead to meaningful extrapolations, at least in terms of upper bounds. For example, a curve π is said to be starshaped on the interval $[0,x_1]$, if for every $x_0 \le x$ in $[0,x_1]$, $\pi(x_0) \le x_0(\pi(x)-\pi(0))/x$. In this paper we will assume that $\pi(0)$ is known, and for convenience, that it is 0. Then, if π is starshaped on $[0,x_1]$, and U is an upper confidence bound (UCB) for $\pi(x_1)$, then x_0U/x_1 is a UCB for $\pi(x_0)$. (Of course this procedure is actually *interpolation*, not extrapolation.) Even if it doesn't establish safety, such a UCB provides an assessment of the state of knowledge and uncertainty about the test substance at dose level x_0, for comparisons with standards and to motivate further study.

Certain assumptions about the curvature of dose-response curves seem to be physically plausible. One such assumption is sigmoidicity. A dose-response curve π is said to be sigmoidal if it is nondecreasing, and there is a point M to the left of which π is convex and to the right of which π is concave. For convenience we'll also assume that π is differentiable at M. Then, if π is sigmoidal, it has a left (or right) derivative π' that is unimodal and maximized at M. (See Royden (1968, Proposition 5.16).) Typically also $\pi(x) \longrightarrow 1$ as $x \longrightarrow \infty$, in which case π is a distribution function, and then π' is a unimodal density function with mode M. In analogy with distribution functions we call the point M the mode.

If π is sigmoidal, then the response-to-dose ratio ρ, defined by $\rho(x) = \pi(x)/x$, is also unimodal, and is maximized at a point $H \geq M$. This can be seen geometrically, or proved as follows: Observe that $\rho' = (x\pi'(x)-\pi(x))/x^2$. If π' is unimodal and maximized at M, then for $x \leq M$,

$$\pi(x) = \int_0^x \pi'(t)dt \leq \int_0^x \pi'(x)dt = x\pi'(x).$$

That is, $x\pi'(x)-\pi(x) \geq 0$. Let

$$H = \inf\{x : x\pi'(x)-\pi(x) < 0\}.$$

Then $H \geq M$ and, for $x \leq H$, $x\pi'(x)-\pi(x) \geq 0$. For $x \geq H$,

$$\begin{aligned}\pi(x) &= \pi(H) + \int_H^x \pi'(t)dt \\ &\geq \pi(H) + \int_H^x \pi'(x)dt = \pi(H) + \pi'(x)(x-H).\end{aligned}$$

Thus for $x \geq H$, $x\pi'(x)-\pi(x) \leq H\pi'(x)-\pi(H) \leq H\pi'(H)-\pi(H) \leq 0$.

Since $H \geq M$, we will call H the *hypermode*, and if ρ is unimodal, we will say that π is *unihypermodal* (even if π isn't sigmoidal). If π is unihypermodal with hypermode H, and $x_0 \leq x \leq H$, then a UCB for $\pi(x_0)$ can be obtained from a UCB for x, since π is then starshaped on $[0,H]$. Unfortunately, H is almost never known in practice.

In Section 2 of this paper we discuss a computational algorithm for ML estimation subject to unihypermodality. In Section 3 we discuss testing hypotheses about the location of the hypermode H. Two classes of tests are discussed, one based on a likelihood ratio statistic, and the other on a weighted sum of contrasts. The likelihood ratio statistic is computed using the maximum likelihood estimate (MLE) subject to unihypermodality. The hypothesis tests lead to a lower confidence bound (LCB) for H, which in turn can be used to construct a UCB for $\pi(x_0)$.

The algorithm for computing the unihypermodal MLE has guaranteed convergence. This does not appear to be manageable under sigmoidicity. (See Schmoyer (1984).) The estimate itself, however, does not provide the kind of smoothing that is the appealing feature of the sigmoid estimate. This is true for two reasons. First, on (H,∞), π can be decreasing. Second, since π' need not be unimodal, π itself can be somewhat irregular. The first problem can be circumvented by assuming that π is both monotone and unihypermodal, although doing this introduces complications in practice. The second difficulty seems unavoidable without the stronger assumption of sigmoidicity. Although the MLE is easy to compute, our primary reason for considering it will be likelihood ratios, rather than the estimate itself.

2. MAXIMUM LIKELIHOOD ESTIMATION SUBJECT TO UNIHYPERMODALITY.

Suppose that dose levels $x_1,\cdots,x_k$ of a substance are administered to $n_1,\cdots,n_k$ subjects, of which $a_1,\cdots,a_k$ respond in a specified way. Let H_i denote the set of all vectors $p = (p_1,\cdots,p_k)'$ for which $0 \le p_j \le 1$ for $j = 1,\cdots,k$, and $p_1/x_1 \le \cdots \le p_i/x_i \ge \cdots \ge$

p_k/x_k. Also let $H = H_1 \cup \cdots \cup H_k$. Then there is a unihypermodal dose-response curve π for which $p_j = \pi(x_j)$, for $j = 1,\cdots,k$, if and only if $p \in H$.

The constraints $p_1/x_1 \leq \cdots \leq p_i/x_i$ might be appropriately called "starshaped restrictions," but differ from the starshaped restrictions considered, for example, in Shaked (1979) or Dykstra and Robertson (1982). The restrictions considered there and here are both induced by the starshapedness of a distribution function, but there the parameters correspond to values of the density of the distribution rather than the distribution function itself, as is the case here. The density of a starshaped distribution function is said to be *increasing on the average*, as are the corresponding parameters. Here, the parameters $p_1,\cdots,p_i$ must satisfy the stronger constraints induced directly by the constraints on π rather than π'.

In many applications, ML estimation in the absence of a concomitant confidence theory is not very useful. Nevertheless there are instances where the best possible point estimate is desired, for example, where it is desired to reduce the data from each of several bioassays in order to analyze them collectively, or in the design of a subsequent study intended, for example, to estimate the response probability at a single well-chosen dose. Another application is data smoothing for graphics. Finally, MLE's are used to compute likelihood ratio statistics, as in the next section.

Thus we consider ML estimation subject to unihypermodality; we seek to maximize the log-likelihood

$$\sum_{i=1}^{k} (a_i \log(p_i) + (n_i - a_i)\log(1-p_i)) \tag{2.1}$$

subject to $p \in H$. We will maximize (2.1) on H by maximizing it on each H_i individually, and then finding the overall maximum.

The set H_i defines a particular kind of *rooted tree order*, which is a kind of partial order. (See Barlow, Bartholomew, Bremner, and Brunk (1972, Chapter 2).) Using the terminology of Barlow et al., in this order an *immediate predecessor* of an element x_j is x_{j+1} if $j < i$, and x_{j-1} if $j > i$. The element x_i, which is called the *root*, has no immediate predecessor. The feature of a tree order that distinguishes it from an arbitrary partial order is that except for the root every element has exactly one immediate predecessor. For given $p_1,\cdots,p_k$, the *value* of an element x_j is p_j/x_j. An element x_j will be called a *violator* if its value exceeds that of its immediate predecessor.

We will say that a block of elements $x_\ell,\cdots,x_m$ is *pooled* if $p_\ell = s'x_j$ for $j = \ell,\cdots,m$, where s' maximizes (in s)

$$\Lambda(s) = \sum_{j=\ell}^{m} [a_j \log(sx_j) - (n_j-a_j)\log(1-sx_j)], \tag{2.2}$$

subject to $0 \le s \le 1/x_m$. The quantity s' is then the value of the block. It is easy to show that (2.2) is a concave function of s on $[0,1/x_m]$. Therefore, a search technique, for example a Fibonacci search (see Zangwill (1969, p.121)), can be used to maximize (2.2) to any desired degree of accuracy (in s) in a finite number of steps. Alternatively (2.2) can be maximized by finding a zero of its first derivative, using Newton's method. Of course, Newton's method tends to be considerably faster in practice than the Fibonnaci search.

To determine the ML partition into blocks and the MLE we will use a maximum-violator algorithm (Thompson (1962), and Thompson and Moore

(1963)). Starting with the sample proportions $\hat{p}$ ($\hat{p}_j = a_j/n_j$ for $j = 1,\cdots,k$) the algorithm searches for the maximum violator in a tentative estimate p. The maximum-violator algorithm then pools the block of elements containing the maximum violator with the block containing its immediate predecessor, as described in (2.2). After pooling, the search for a maximum violator is started over, and the procedure continues until there are no more violations.

Thompson (1962) shows that the maximum-violator algorithm can be used to determine isotonic (least squares) regression estimates subject to a rooted tree order. Robertson and Wright (1980) prove that a minimum-lower-sets algorithm yields MLE's in a setting that is much more general than least-squares. A minimum-lower-sets algorithm could be used to compute the MLE in the present problem, but would not take advantage of the tree order.

A pooling scheme is said to have the Cauchy mean value property if the value of a block formed by pooling two previous blocks is bounded by the previous blocks' values. (See Robertson and Wright (1980).) In order to prove that the maximum-violator algorithm terminates at the MLE, we will use the following, which actually holds for any pooling scheme having the Cauchy mean value property and the rooted tree order defined above:

LEMMA. *Suppose a block containing a violator is pooled with the block containing its immediate predecessor. If* MVB *denotes the value of the maximum violator before pooling, and* MVA *is the value of the maximum violator after pooling, then* MVB $\geq$ MVA.

PROOF. Suppose that v is a violator, ip its immediate

predecessor, w the first element in the block (if it exists) that is immediately preceded by the block containing v. The value of w is not changed by pooling ip and v. If w is a violator (before pooling), then its value does not exceed that of v, hence its value does not exceed MVB.

Because Λ' is strictly decreasing, it is easy to show that the pooling (2.2) has the Cauchy mean value property. Suppose that ip is not contained in the root block, and let ip′ be the element closest to the root in the block containing ip. Then upon pooling ip and v, ip′, which besides w is the only element that could possibly become a new violator, takes on a new value that is bounded above by the original value of v, hence is bounded above by MVB. Suppose on the other hand that ip is contained in the root block. Let h be the first element in the block (if it exists) that is preceded by the root block, but does not contain v. Then the value of h, which is the only element other than w that could become a new violator after pooling ip and v, is unchanged by the pooling. If h is a violator (before pooling), then its value does not exceed MVB. If h is not a violator (before pooling), then its value does not exceed that of ip, which is exceeded by the value of v and hence also by MVB. Thus MVB $\geq$ MVA.

Let $\tilde{p}^i = (\tilde{p}^i_1, \cdots, \tilde{p}^i_k)'$ be the MLE of p subject to $p \in H_i$. Then

$$(2.3) \qquad \tilde{p}^i_1/x_1 \leq \cdots \leq \tilde{p}^i_i/x_i \cdots \tilde{p}^i_k/x_k .$$

Because (2.1) is a (pseudo) concave function of $p_1, \cdots, p_k$, where there is a strict inequality in (2.3), the corresponding constraint is inactive and can be ignored in the maximization. Then, the log-likelihood splits into separate sums that are maximized independently. Elements satisfying $\tilde{p}^i_\ell/x_\ell = \cdots = \tilde{p}^i_m/x_m$ of $\tilde{p}^i$ are pooled as in (2.2). Because pooling two blocks never increases the likelihood, there is no division of an ML block into sub-blocks that have different values, and for which there is no violation. From this it is clear that each ML block can be formed by starting with $\hat{p}$, iteratively pooling violators, and stopping when there are no violators. The problem is to determine which violators to pool.

Consider a sequence of these poolings, all of which are correct in the sense of being involved in the final ML partition into blocks, and suppose that at a given stage of the sequence a particular element m is a maximum violator. Let ipm denote the immediate predecessor of m. We distinguish four types of such poolings, at least one of which must occur if the m-ipm violation is to be eliminated: (1) m can be pooled with ipm; (2) ipm, if it does not belong to the root block, can be pooled with elements preceding it; (3) ipm, if it belongs to the root block, can be pooled with elements in the block (if any) preceded by the root block but not containing m; and (4) m can be pooled with elements that it precedes (if any).

Suppose that a type (1) pooling is not involved in the final MLE. Because poolings are restricted to those involving a violator and immediate predecessor, poolings of type (4) can only increase the value of m. Similarly, poolings of type (2) can only decrease the value of ipm. Thus, only poolings of type (3) could remove the violation. However, if a type (3) pooling elevates the value of ipm to or above that of m, it would itself have to involve a violator having value greater than that of m. By the lemma, m could not be a maximum violator at the given stage.

Since pooling never increases the likelihood, it is now clear that if the maximum-violator algorithm starts at $\hat{p}$, then it terminates at the MLE. Since maximizing (2.2) can be accomplished with guaranteed convergence, the same is true about the algorithm as a whole.

Example 2.1. The data in Table 1 is from an experiment performed by Dalbey and Lock [1982] to assess the toxicity of diesel fuel aerosol (DFA) smoke screens. Rats were enclosed in chambers in which the level of DFA could be monitored and controlled. Table 2 contains the unihypermodal MLE computed from this data. For reference, the sample proportions and monotone and sigmoid MLE (see Schmoyer (1984)) are also given. The unihypermodal MLE, it turns out, is in H_6. The unihypermodal MLE is illustrated in Figure 1, along with the sigmoid estimates and sample proportions. It is clear from the figure that the unihypermodal estimate does not provide the same degree of smoothing as the sigmoid estimate. Assuming monotonicity would not improve it, since it is already monotone.

Table 1. Results of DFA Experiment.

Dose h·mg/ℓ	Number of trials	Proportion that died
8.0	30	0
16.0	40	0.025
24.0	40	0.05
28.0	10	0.5
32.0	30	0.4
48.0	20	0.8
64.0	10	0.6
72.0	10	1

Table.2. Maximum Likelihood Estimates for DFA Study.

Dose (h•mg/ℓ)	Unconstrained	Monotone	Sigmoid	Unihypermodal
8.0	0	0	0	0
16.0	0.025	0.025	0.025	0.025
24.0	0.05	0.05	0.05	0.05
28.0	0.5	0.425	0.390	0.382
32.0	0.4	0.425	0.448	0.436
48.0	0.8	0.733	0.677	0.8
64.0	0.6	0.733	0.892	0.8
72.0	1	1	1	0.9

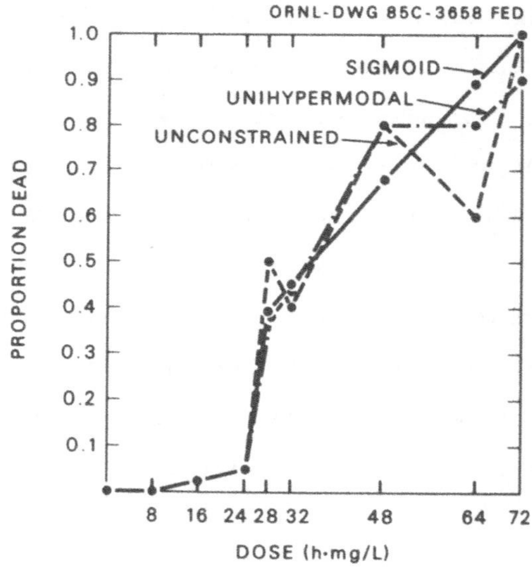

Figure 1. Unconstrained, Sigmoid and Monotone Maximum Likelihood Estimates for DFA Study.

The unihypermodal MLE was computed using a SAS program (available on request), and took about 2.6 seconds of CPU time to compute. Computing the sigmoid MLE does not require substantially more time. However, the algorithm for computing the sigmoid MLE, suggested by Schmoyer (1984), is not guaranteed to converge, and can fail in practice, as Schmoyer demonstrates. Thus, while most people would probably prefer the sigmoid MLE to the unihypermodal MLE on the basis of fundamental notions of smoothness, the relative ease of computation of the latter may make it preferable even as a point estimator in some applications.

3. LOWER CONFIDENCE BOUNDS FOR H. An LCB for H is the smallest acceptable H' in tests of the hypotheses $H \le H'$ against $H > H'$. If an LCB corresponds to a test having good power characteristics, then the LCB itself is generally considered to be good. Let S denote the set of all p satisfying the sigmoidicity constraints (see Schmoyer (1984)), and let M denote the set of all p satisfying monotonicity (increasing) constraints. For $j = 1,\cdots,k$, let $N_i = [H_1 \cup \cdots \cup H_i] \cap S$. It is not difficult to see that $\{p \in S : p_j = \pi(x_j), j = 1,\cdots,k, H \le H'\} = N_i$, where $i = \min\{j : H' \le x_j\}$. Perhaps the ideal test of $H \le H'$ would be based on the likelihood ratio (LR), with numerator likelihood maximized subject to $p \in N_i$ and denominator likelihood maximized subject to $p \in S$. Such a test could be performed by performing individual LR tests of $H_j \cap S$ against S, $j = 1,\cdots,k$. Unfortunately, the problem of determining a p-value or least favorable configuration (LFC) in this setting seems to be analytically intractable, and is sufficiently complex that search techniques appear to be unfeasible as well. This is true even if S is replaced with one of the weaker alternatives M, H, or $M \cap H$.

In order to obtain a tractable testing procedure, we need to modify the hypothesis being tested, or the test, or both in an appropriate way. In this regard we will consider two approaches. The first is based on the asymptotic conditional chi-square distribution of LR. The second approach abandons LR completely in favor of a class of linear test statistics. The LCB's for H that we so obtain will then be lower than that achievable if a fully efficient procedure could be used. Of course, in this setting "lower" translates to "on the safe side."

3.1 *Conditional Chi-Square Tests.* Let U denote $\{p : 0 \le p_j \le 1,\ j = 1,\cdots,k\}$, and let LR^i denote the LR for testing $p \in H_i$ against U. The distributions of the LR^i are asymptotically conditionally chi-square, with degrees of freedom equal to the number of constraints actively bounding the corresponding MLE's. This follows from Hartigan (1967, Theorem 1) and (1) that $\hat{p}$ is asymptotically $N(p,D)$, where D is the diagonal matrix with diagonal elements $p_j(1-p_j)/n_j$ for $j = 1,\cdots,k$, (2) that the MLE is asymptotically either $\hat{p}$ or the D-weighted least squares projection of $\hat{p}$ onto the intersection of one or more of the linear spaces that define the null hypothesis, and (3) that the possible subspaces onto which the MLE is a projection asymptotically all contain p. Thus an asymptotic size α test can be devised from conditional size α tests using the MLE subject to H_i. Since the asymptotic approximation used here will be good only if (3) is a good approximation, the $\hat{p}_j$ must have fairly small variances in addition to approximate normality.

If the variance-covariance matrix of $\hat{p}$ were translation invariant, then by (1) and (2) the conditional chi-square tests based on the LR^i would still be asymptotically size α even without (3) (Hartigan (1967, Theorem 2)). Unfortunately, variance stabilizing transformations, for example, $\arcsin\sqrt{p}$, destroy the linear structure of the null hypothesis. Furthermore, monotone nonlinear transformations of π such as $\arcsin \sqrt{\pi}$ do not preserve features like unihypermodality or sigmoidicity, as may easily be shown. Thus it seems that Hartigan's Theorem 2 cannot be applied.

In addition to a strong dependence on asymptotics, the tests based on the LR^i are deficient in their alternative, which is U (rather than S). The tests are therefore sensitive to deviations from H, which of course is assumed. This also forces the choice of null hypothesis, H_i rather than $H_i \cap S$. Otherwise the tests would be sensitive to deviations from S. Another disadvantage of conditional chi-square tests is that the number of constraints that actively bound the MLE—information that bears on the null hypothesis—is lost in conditioning. Finally, there is also a problem in what to do when the number of active constraints turns out to be zero. In that case it would seem unreasonable to reject, though the conditional approach dictates rejection 100α percent of the time (forcing randomization). Unfortunately there does not seem to be any clearly superior alternative to the conditional LR tests, and so we will consider them as a means of obtaining an LCB for H in spite of their disadvantages.

3.2 *Linear Tests*. As an alternative to the LR approach, we also consider tests based on linear statistics. Linear test statistics are considered for testing linear (equality constrained) null hypotheses about normal means against order restricted alternatives by Abelson and Tukey (1963) and more generally by Schaafsma and Smid (1966) and Schaafsma (1968). The advantages and disadvantages of these tests in this context are discussed in Barlow et al. (1972, Chapter 4). Tractability is a strongpoint. In the present context, the inequality-constrained null hypothesis would seem less suited to linear test statistics. The advantage of tractability on the other hand will be come clear.

For $i = 1,\cdots,k-1$, consider testing $p \in N_i$ against $\{p \in S : p_j = \pi(x_j),\ j = 1,\cdots,k,\ H > x_i\}$, and restrict attention to linear test statistics of the form

$$T_i = \sum_{j=i}^{k-1} w_j \left[\frac{\hat{p}_{j+1}}{x_{j+1}} - \frac{\hat{p}_j}{x_j}\right], \tag{3.1}$$

where the weights w_j, defined for $j = i,\cdots,k-1$, must be nonnegative and sum to 1. The value of $\delta = E_p(T_w)$ is clear from (3.1). Under the null hypothesis $p_i/x_i \geq \cdots \geq p_k/x_k$, and thus $\delta \leq 0$, and T_w should tend to be negative. Under the alternative, $H > x_i$ and the expectation of T_w consists of a series (possibly none) of nonnegative terms followed by a series of nonpositive terms. The farther H is from x_i, the greater will be the number of nonnegative terms, and the larger δ will be. From this it is clear that T_w should reject N_i when it is too large.

Unfortunately, $\delta \leq 0$ is possible under the alternative, particularly if the w_j are relatively large for j such that $x_j > H$. This problem will have to be reckoned with when choosing the weights. However, because T_w is linear in $\hat{p}$, its distribution should be well suited to asymptotic approximation. As will be shown, T_w also has the advantage of being tractable enough to admit the computation of asymptotic LFC's. Although relying on the asymptotic least favorability of a particular point does introduce into the method an additional degree of dependence on asymptotics, that dependence is not as strong as it is in the conditional chi-square approach, simply because the true configuration is not likely to be as unfavorable as the asymptotic LFC. In the conditional approach the strong dependency on asymptotics is always present, regardless of the configuration. For these reasons we consider T_w, in spite of the fact that the quantity δ incompletely characterizes null and alternative values of p. It is easy to show that

$$T_w = \sum_{j=i+1}^{k-1} (w_j - w_{j-1})\hat{p}_j/x_j - w_i p_i/x_i + w_{k-1}\hat{p}_k/x_k.$$

It follows that

$$(3.2) \qquad \mathrm{Var}_p(T_w) = \sum_{j=i+1}^{k-1} (w_j - w_{j-1})^2 V_j + w_i^2 V_i + w_{k-1}^2 V_k,$$

where $V_j = \mathrm{Var}(\hat{p}_j)/x_j^2 = p_j(1-p_j)/(n_j x_j^2)$, for $j = 1,\cdots,k$. Let $N = \sum_{j=1}^{k} n_j$, and suppose that N approaches infinity in such a way that for $j = 1,\cdots,k$, k fixed, n_j/N converges to a limit c_j strictly between 0 and 1. Let $\langle T_w^N, t_w^N\rangle$ denote a sequence of size α tests, with test statistics T_w^N of the form (3.1) and critical points t_w^N. Also let $\delta = E_p(T_w^N)$ be the same for every N.

Since T_w^N converges in probability to δ, if $\delta^* = E_{p*(w)}(T_w^N)$ where $p^*(w)$ is an asymptotically LFC for w, then $\delta^* = 0$. In addition, for any p, $(T_w^N - \delta)/\mathrm{Var}_p(T_w^N)^{1/2}$ has in the limit, the distribution of a standard normal random variable Z. Thus $t_w^N/\mathrm{Var}_{p*(w)}(T_w^N)^{1/2} \longrightarrow Z^{1-\alpha}$, and $p^*(w)$ maximizes $\mathrm{Var}_p(T_w^N)$ (or $N\cdot\mathrm{Var}_p(T_w^N)$) subject to $p \in N_i$ and $E_p(T_w^N) = 0$ if and only if

$p_j/x_j = p_{j+1}/x_{j+1}$ for all j such that $w_j > 0$. If $w_j > 0$ for $j = i,\cdots,k'-1$, and $w_j = 0$ for $j = k',\cdots,k-1$, this is equivalent to $p_j/x_j = s$, say, for $j = i,\cdots,k'$. Then (3.2) becomes

$$\text{(3.3)} \qquad \mathrm{Var}(T_w^N) = \sum_{j=i+1}^{k'-1} (w_j-w_{j-1})^2 s(1-sx_j)/n_jx_j$$
$$+ w_i^2 s(1-sx_i)/n_ix_i + w_{k'-1}^2 s(1-sx_{k'})n_{k'}x_{k'},$$

and an asymptotic least favorable value of s (and corresponding LFC) may be found by straightforward differentiation of (3.3) subject to $s \le 1/x_{k'+1}$.

We now have a class of asymptotically size α tests $\langle T_w^N, t_w^N\rangle$ indexed by w. The question becomes how to choose w. We proceed as follows: In order to mitigate the effects of δ possibly being negative under the alternative, we choose two numbers k_0 and i_0, and take $w_j = 0$ for $j > k_{max} = \max(k_0-1, i+i_0)$. For sufficiently large values of i the only nonzero weights are then $w_i,\cdots,w_{i+i_0}$ and δ can be negative under the alternative only for the few values of i for which x_i is sufficiently close to H. For small values of i, if the only nonzero weights were $w_i,\cdots,w_{i+i_0}$, then a test based on T_w could fail to reject simply because $\hat{p}_i,\cdots\hat{p}_{i+i_0+1}$ are all small. If x_{k_0} is reasonably close to H, however, then this latter problem is circumvented by allowing $w_i,\cdots,w_{k_0-1}$ all to be nonzero. Of course if x_{k_0} is much less than H, the problem can still occur. If x_{k_0} exceeds H appreciably, then H could exceed x_i appreciably with δ remaining negative. In practice k_0 should be guessed, possibly on the basis of a prior (e.g., range-finding) study, with the intent of making x_{k_C} close to H. A poor guess may result in unpowerful tests, but the tests are of course asymtotically valid in any case.

We now choose exact values for $w_i, \cdots, w_{k-1}$ in accordance with the restrictions just described. Naturally we seek to maximize power under the alternative hypothesis. However, there are alternatives for which $\delta \le 0$. Also, for any fixed p_0 for which $\delta > 0$, $P_{p_0}(T^N > t^N) \longrightarrow 1$ as $N \longrightarrow \infty$. Therefore, we restrict attention to sequences of alternatives $\langle p_N \rangle$ for which $\delta_N = E_{p_N}(T_w^N) > 0$, and which converge to a limit p in N_i. We also require that $\delta_N \longrightarrow 0$ at an appropriate rate, for example that for some $\epsilon > 0$,

$$(3.4) \qquad \delta_N = t_w^N + \frac{\epsilon}{N^{1/2}}.$$

It follows that $P_{p_N}(T_w^N > t_w^N)$ is asymptotically equivalent to

$$(3.5) \qquad P\left[Z > \frac{-\epsilon}{[N\cdot \mathrm{Var}_p(T_w^N)]^{1/2}}\right],$$

where Z is standard normal. In the limit, (3.5) is minimized in p when $N\cdot\mathrm{Var}_p(T_w^N)$ is maximized. Since $p \in N_i$, this occurs at $p^*(w)$. Therefore, the minimum asymptotic power among all sequences of alternatives satisfying (3.4) and convergent to a limit in N_i is maximized in w when $N\cdot\mathrm{Var}_{p*(w)}(T_w^N)$ is minimized. The minimizing w, call it w^{**}, along with its corresponding LFC, $p^{**} = p^*(w^{**})$, comprise a saddlepoint of $\mathrm{Var}_p(T_w^N)$ on $\{w, p : p \in N_i, \delta = 0, w \ge 0, 1'w = 1, \text{ and } w_j = 0 \text{ for } j > k_{max}\}$. The test corresponding to w^{**} is maximin in this sense.

In order to determine the saddlepoint (w^{**}, p^{**}), we next consider the problem of minimizing $Var_p(T_w^N)$ in w, subject to $w \geq 0$, $1'w = 1$, and $w_j = 0$ for $j > k_{max}$, where we assume that $p_j = sx_j$ for $j = 1, \cdots, k_{max}+1$, and $s \leq 1/x_{k_{max}+1}$. Then $Var_p(T_w^N)$ can be written as in (3.3), and by (3.3) can be expressed as $w'\Sigma w$, where the nonnegative definite matrix Σ is defined below. The minimization, then, becomes a quadratic programming (QP) problem, and could be handled using a general QP approach. However, we now show that if (3.3) is minimized subject to $1'w = 1$ alone, then that minimum satisfies $w \geq 0$ automatically.

Using a Lagrange multiplier λ, the minimum of $w'\Sigma w$ subject to $1'w = 1$ occurs where $2\Sigma w = \lambda 1$. To find this minimum then we need to solve $\Sigma w = 1$. It is straightforward to show that the matrix $\Sigma = (\sigma_{\ell m})$ has the form

$$\sigma_{\ell m} = \begin{cases} V_{\ell+i-1} + V_{\ell+i} & \text{if } |\ell - m| = 0 \\ -V_{\min(\ell,m)+i} & \text{if } |\ell - m| = 1, \\ 0 & \text{if } |\ell - m| > 1 \end{cases}$$

for $\ell, m = 1, \cdots, k-i$.

By Farkas' Lemma (see Zangwill (1969)) there is a $w^* \geq 0$ such that $\Sigma w^* = 1$ if and only if $\Sigma w \geq 0$ implies $1'w \geq 0$. The ℓth component of Σw is $-V_{\ell+i-1}w_{\ell+i-2} + (V_{\ell+i-1}+V_{\ell+i})w_{\ell+i-1} - V_{\ell+i}w_{\ell+i}$, where of course $w_{i-1} = 0$ and $w_j = 0$ for $j > k_{max}$. Suppose that $1'w < 0$. Let w_m be a most negative component of w (possibly among several). We may assume that $s > 0$, since the point $p = 0$ is not an LFC for any admissible set of weights. Since $s \leq 1/x_{k_{max}+1}$, this implies $V_\ell > 0$ for $\ell = i, \cdots, k_{max}$. If all components of w are equal, then $(V_i+V_{i+1})w_i-X_{i+1}w_{i+1} < 0$. Otherwise we can choose w_m so that either $w_m-w_{m-1} < 0$ or $w_m-w_{m+1} < 0$, and thus $-V_m w_{m-1} + (V_m+V_{m+1})w_m - V_{m+1}w_{m+1}$, which is $V_m(w_m-w_{m-1}) + V_{m+1}(w_m-w_{m+1})$, is negative. Therefore it is false that $\Sigma w \geq 0$. This shows that there is a $w^* \geq 0$ satisfying $\Sigma w^* = 1$.

If w is fixed, $w \geq 0$, $1'w = 1$, and $w_j > 0$ if and only if $j = i$ or $\cdots$ or k_{max}, then we can find a $p \in N_i$ with $\delta = 0$ that maximizes $Var_p(T_w^N)$. Now consider alternately solving the maximization and minimization problems, starting, for example, with equality among the nonzero weights. If the algorithm terminates at a fixed point (w^{**}, p^{**}), where $w_j^{**} > 0$ for $j = i, \cdots, k_{max}$, then that point is up to machine accuracy the desired maximin set of weights and corresponding asymptotic LFC. We have written a SAS program to do this. Although we have not proved that the algorithm always converges, that has been our experience.

Example 2.1 (continued). Table 3 contains the results of applying the conditional chi-square tests to the DFA data, discussed in Section 2. Table 4 contains the results for the linear tests, including the optimal weights. The value taken for k_0 is 6, and the value of i_0 is 2. On the basis of either approach the 95% LCB for H is 24.

The greatest significance level for the conditional tests is only .12, suggesting possible lack of fit. It is possible that the conditional chi-square tests could have rejected for every $i = 1, \cdots, k$. That event should be construed as evidence of lack of fit of the model S, and rejecting S on that basis would be an anti-conservative procedure, as may easily be shown. However, it would not be a good

Table 3. Results of Conditional Chi-Square Tests of $p \in H_i$, $i = 1,\cdots,k$, for DFA Study.

i	log-likelihood	df = number of active constraints	-2log(LR)	p-value
1	-76.95	7	40.95	< .0001
2	-74.11	6	35.26	< .0001
3	-68.89	5	24.82	.0002
4	-59.30	2	5.64	.06
5	-59.40	3	5.84	.12
6	-58.95	2	4.94	.08
7	-60.37	4	7.79	.10
8	-59.80	3	6.65	.08

Table 4. Results of Linear Tests $(k_0=6, i_0=2)$ of $p \in N_i$, $i = 1,\cdots,k-1$, for DFA Study.

i	w_1	w_2	w_3	w_4	w_5	w_6	w_7	$Var(T_w) \cdot 10^{-6}$	$T_w \cdot (Var(T_w))^{1/2}$	p-value
1	.049	.157	.271	.284	.238	0	0	.743	4.54	< .0001
2	0	.139	.290	.308	.264	0	0	.906	4.30	< .0001
3	0	0	.308	.354	.338	0	0	1.53	3.52	.0002
4	0	0	0	.316	.684	0	0	6.40	.459	.32
5	0	0	0	0	.484	.516	0	3.17	-.98	.84
6	0	0	0	0	0	.493	.507	2.46	-.84	.80
7	0	0	0	0	0	0	1	10.9	1.37	.09

way to test goodness of fit (GOF). To be on the safe side, the null hypothesis in a GOF test should be that the model S does *not* fit the data. Rejecting the null hypothesis would then constitute statistical proof that the model is valid. Unfortunately, the complement of S in U or M is a very intractable set. GOF testing will not be considered further here.

Of course the choice among the conditional chi-square, the linear, and any other approach to testing hypotheses about the location of H must be made a priori, as must the choice of k_0 and i_0 if the linear approach is chosen. It seems unlikely that any approach would be uniformly optimal over all designs $x_1,\cdots,x_k$ and $n_1,\cdots,n_k$, or all $p_1,\cdots,p_k$. The very great variety of possible designs and values of p in this setting would make a general simulation study of which method to choose an immense effort. Instead we suggest basing the choice on a prior simulation study at the particular design under consideration, and with the value of p guessed, or chosen as a "worst" case, or both.

To illustrate this idea we will use the sigmoid MLE and the concave MLE, (.11, .22, .33, .39, .44, .67, .89, 1) (Schmoyer (1984)), as choices of p. The value of H for the linearly interpolated sigmoid MLE is 48, although a slight change in p would result in H = 28. Although the linearly interpolated concave MLE satisfies H = 72, it is also the limit of a sequence of concave curves, each satisfying H = 0. The concave MLE can be considered a "worst" case since under strict concavity of π, linear interpolation between the origin and $(x,\pi(x))$ leads to lower (not upper) bounds on π, and a nonstrictly concave curve is the most difficult type of concave curve to detect. One thousand simulation trials were run for both values of p for the conditional chi-square tests and the linear tests with $i_0 = 2$ and $k_0 - 4,\cdots,8$.

The frequencies of the various LCB's for H are given in Table 5 (sigmoid MLE) and Table 6 (concave MLE). Evidently, the conditional chi-square and the linear tests, with the exception of the case

$k_0 = 4$, perform similarly and acceptably well at both the sigmoid and concave MLE's. The linear tests appear to be slightly conservative. If the experimental results are to be used in decisions about safety, slightly conservative procedures should be preferred to procedures that are slightly anti-conservative. Nevertheless, on the basis of these limited simulation results, for these particular design conditions, the conditional chi-square approach would seem preferable, especially since it does not require choosing quantities such as k_0 or i_0.

Table 5. 95% LCB Frequencies in 1000 Simulation Trials with p = Sigmoid MLE.

	Conditional			Linear		
LCB	Chi-square	k_0=4	k_0=5	k_0=6	k_0=7	k_0=8
0	0	542	15	2	0	0
8	2	6	15	1	0	0
16	15	42	47	39	28	15
24	933	410	884	908	936	959
28	29	0	37	44	25	18
32	12	0	2	6	11	8
48	1	0	0	0	0	0
64	0	0	0	0	0	0
72	8	0	0	0	0	0

Suppose that a UCB for $p_0 = \pi(x_0)$ is desired, where $x_0 < x_1$ is a dose level of interest. If particular $x_{m_1},\cdots,x_{m_\ell}$ are chosen a priori, where $x_{m_\ell} \leq H$, then a valid UCB for p_0 can be obtained from the data at those doses, and the LCB for H provides a check on

Table 6. 95% LCB Frequencies in 1000 Simulation Trials with p = Concave MLE.

	Conditional	Linear				
LCB	Chi-Square	$k_0=4$	$k_0=5$	$k_0=6$	$k_0=7$	$k_0=8$
0	942	963	962	955	953	962
8	26	14	10	20	20	12
16	12	22	11	11	19	18
24	5	1	17	10	7	6
28	6	0	0	4	1	1
32	2	0	0	0	0	0
48	1	0	0	0	0	1
64	2	0	0	0	0	0
72	4	0	0	0	0	0

the assumption $x_{m_\ell} \le H$. For example, for some confidence level $1-\alpha$, and $\alpha^* = 1-(1-\alpha)^{1/\ell}$, let $U_j^{1-\alpha^*}$ denote the usual level $1-\alpha^*$ UCB for the response probability p_j, based solely on the data obtained at x_j. Then

$$U_H(x_0) = \min_{j \le \ell} U_{m_j}^{1-\alpha^*} x_0/x_j$$

is a $1-\alpha$ UCB for $\pi(x_0)$, as may easily be shown. Confidence bounds similar to $U_H(x_0)$ (but not involving H) are discussed in Korn (1982). Unfortunately, H is almost never known a priori.

Now suppose that H is unknown, and let L denote the above asymptotic LCB for H. Let $1-\alpha'$ denote the actual confidence level of L. Define j_L to be the smaller of that j for which $x_j = L$ and some integer upper bound j_B. For predefined α, let $\alpha^* = 1-(1-\alpha)^{1/j_B}$. If $j_L \ne 0$, take

$$U_L(x_0) = \min_{j \le j_L} U_j^{1-\alpha^*} x_0/x_j .$$

If $j_L = 0$, take $U(x_0) = 1$. Then

$$P[p_0 \le U_L(x_0)] \ge P(p_j \le U_j^{1-\alpha^*},\ j = 1,\cdots,j_B \quad L \le H)$$

$$\ge 1-[1-P(p_j \le U_j^{1-\alpha^*},\ j = 1,\cdots,j_B]-[1-P(L \le H)] = 1-\alpha-\alpha'.$$

In this example, suppose that a priori we had taken $\alpha = \alpha' = .025$, and $j_B = 4$. (Of course, this choice of α and α' could be entirely inappropriate.) Then $\alpha^* = .0063$, $L = 24$ and $j_L = 3$, and $U_1 = .1553$, $U_2 = .1660$, and $U_3 = .2059$. Therefore $U_1/x_1 = .0194$, $U_2/x_2 = .0104$, and $U_3/x_3 = .0086$, and hence the 95% $U_L(x_0) = .0086x_0$. As men receive field exposure concentrations of DFA in the vicinity of .5 mg/L for several hours, the safety of DFA clearly cannot be concluded on the basis of this analysis. In fact, in view of the proximity of this range to levels that are lethal to rats, it might be unreasonable to expect a tighter confidence bound. Further study of DFA appears warranted, and was in fact pursued. Of course if x_0 were much farther from lethal levels, then $U(x_0)$ would be much lower.

Alternatively, suppose that an independent (e.g., subsequent) experiment is performed at dose levels $y_1,\cdots,y_{k'} \le L$, chosen on the basis of the present study. Let U, a function of the second experiment, be a valid $1-\alpha$ UCB for p_0 if $y_{k'} \le H$. Then, once again, since $L \le H$ implies $y_{k'} \le H$, $P(p_0 \le U) \ge 1-\alpha-\alpha'$. The choice of $y_1,\cdots,y_{k'}$ should depend on the apparent shape of π as estimated from the first experiment.

REFERENCES

Abelson, R.P. & Tukey, J. (1963). Efficient Utilization of Non-numerical Information in Quantitative Analysis: General Theory and the Case of Simple Order, *Ann. Math. Statist.* **34**, 1347-1369.

Barlow, R.E., Bartholomew, D.J., Bremner, J.M., & Brunk, H.D. (1972). (1972). *Statistical Inference Under Order Restrictinos*, New York: Wiley.

Dalbey, W.E. & Lock, S. (1972). Inhalation Toxicology of Diesel Fuel Obscurant Aerosol in Sprague-Dawley Rats, ORNL/TM-8867, Oak Ridge National Laboratory, Biology Division.

Dykstra, R.L. & Robertson, T. (1982). Order Restricted Statistical Tests on Multinomial and Poisson Parameters: The Starshaped Restriction, *Ann. Statist.* **4**, 1246-1252.

Hartigan, J.A. (1967). Distribution of the Residual Sum of Squares in Fitting Inequalities, *Biometrika* **54**, 69-84.

Korn, E.L. (1982). Confidence Bands for Isotonic Dose-Response Curves, *Applied Statistics* **31**, 59-63.

Robertson, T. & Wright, F.T. (1980). Algorithms in Order Restricted Inference and the Cauchy Mean Value Property, *Ann. Statist.* **8**, 645-651.

Royden, H.J. (1968). *Real Analysis*, New York: Macmillan.

Schaafsma, W. & Smid, L.J. (1966). Most Stringent Somewhere Most Powerful Tests Against Alternatives Restricted by a Number of Linear Inequalities, *Ann. Math. Statist.* **37**, 1161-1172.

Schaafsma, W. (1968). A Comparison of the Most Stringent and the Most Stringent Somewhere Most Powerful Test for Certain Problems with Restricted Alternative, *Ann. Math. Statist.* **39**, 531-546.

Schmoyer, R.L. (1984). Sigmoidally Constrained Maximum Likelihood Estimation in Quantal Bioassay, *J. Amer. Statist. Assoc.* **79**, 448-453.

Shaked, M. (1979). Estimation of Starshaped Sequences of Poisson and Normal Means, *Ann. Statist.* **7**, 729-741.

Thompson, W.A., Jr. (1962). The Problem of Negative Estimates of Variance Components. *Ann. Math. Statist.* **33**, 273-289.

Thompson, W.A., Jr. & Moore, J.R. (1963). Non-negative Estimates of Variance Components, *Technometrics* **5**, 441-449.

Zangwill, W.I. (1969). *Nonlinear Programming: A Unified Approach*, Englewood Cliffs, New Jersey: Prentice-Hall.

Richard L. Schmoyer
Mathematics & Statistics Research Department
Oak Ridge National Laboratories
Oak Ridge, TN 37831

ON NONPARAMETRIC TESTS FOR ORDERED ALTERNATIVES IN TWO-WAY LAYOUTS

Z. Govindarajulu[1]
University of Kentucky

S. H. Mansouri-Ghiassi
Texas Tech. University

AMS 1980 subject classifications: *Primary* 62G10; *Secondary* 62G20, 62K10.

Key words and phrases: *Asymptotically distribution-free test, ordered alternatives, randomized blocks.*

ABSTRACT

Nonparametric tests for ordered alternatives in two-way layouts are briefly surveyed. An asymptotically distribution-free test is proposed for unequally spaced ordered alternatives in two-way layouts when the ratios of the alternative values of the parameters are known. The test criterion is a linear function of the ranks of the deviates from the estimates of the nuisance parameters. The asymptotic efficiency comparisons relative to the likelihood derivative test and other nonparametric tests for randomized (complete) blocks show that our procedure is generally more powerful.

1. INTRODUCTION.

Let $X_{ijk}\left[1 \le i \le I;\ 1 \le j \le J;\ 1 \le k \le n_{ij};\ n = \sum_{i,j} n_{ij}\right]$ be independent random variables having continuous cumulative distribution functions (c.d.f.'s) $P(X_{ijk} \le x) = F(x-\mu-\beta_i-\tau_j)$

---

1. Research for this paper has been supported in part by the Office of Naval Research contract No. N00014-75-C10003, task order NR042-295, and Grant No. N00014-84-K-D184. Reproduction in whole or in part is permitted for any purpose of the Government of the United States of America.

where μ and $(\beta_1, \cdots, \beta_I)$ are nuisance parameters and $(\tau_1, \cdots, \tau_j)$ are treatment effects. We wish to test

$$H_0 : \tau_1 = \cdots = \tau_J \quad \text{versus} \quad H_1 : \tau_j = \tau a_j \quad (1 \le j \le J)$$

where $a_1 \le \cdots \le a_J$, not all a_j's are equal and $\tau > 0$ is an unknown parameter. Of course F is unknown. This can happen in a real situation, for instance, in an experiment involving replicated regression model, the a_j's (the determinations of the independent variable) are known constants.

Several nonparametric test procedures are available in the literature for testing H_0 vs. H_1 in randomized blocks with $n_{ij} \equiv 1$. Jonckheere (1954) proposed a test criterion based on Kendall's rank correlation coefficient between the postulated order and the observed order.

Let ξ_i denote Kendall's rank correlation coefficient between the postulated order and the observation order in the ith block. Then Jonckheere's (1954) procedure is to reject H_0 for large values of

$$\xi = \sum_{i=1}^{I} \xi_i . \tag{1.1}$$

Page's (1963) procedure rejects H_0 for large values of

$$\rho = \sum_{i=1}^{I} \rho_i \tag{1.2}$$

where ρ_i denotes the Spearman rank correlation coefficient between the postulated order and the observation order in the ith block. Hollander (1967) and Doksum (1967) proposed an asymptotically distribution-free criterion based on the sum of Wilcoxon signed rank statistics (the sum being over all distinct pairs) with $n_{ij} \equiv 1$. Let

$$Y_{uv}^{(i)} = |X_{iu}-X_{iv}|$$

and $R_{uv}^{(i)}$ be the rank of $Y_{uv}^{(i)}$ in the ranking of $Y_{uv}^{(i)}$, $(i = 1,\cdots,I)$. Further let

$$T_{uv} = \sum_{i=1}^{I} R_{uv}^{(i)}\psi_{uv}^{(i)}$$

where

$$\psi_{uv}^{(i)} = 1, \quad \text{if} \quad X_{iu} < X_{iv} \tag{1.3}$$
$$= 0, \quad \text{otherwise.}$$

Then

$$T = \sum_{u<v} T_{uv}. \tag{1.4}$$

Notice that T_{uv} denotes a measure of the difference between the uth and v^{th} treatments and the summation over $u < v$ takes into account the prior ordering of the treatments. We reject H_0 for large values of T. Note that T is not distribution-free because $var(T|H_0)$ depends on F. However, one can obtain a consistent estimate of $var(T|H_0)$.

Doksum (1967) proposed a test procedure that is very similar to the T-statistic. He uses the random variables

$$U_{uv} = T_{uv} - \sum_{i=1}^{I} \psi_{uv}^{(i)}$$

and considers the statistic

$$U = \sum_{u<v}^{J} (U_{u\cdot}-U_{v\cdot}), \quad U_{u\cdot} = \sum_{v=1}^{J} U_{uv}/J.$$

An asymptotically equivalent statistic is

$$T' = \sum_{u<v}^{J} (T_{u\cdot}-T_{v\cdot}), \quad T_{u\cdot} = \sum_{v=1}^{J} T_{uv}/J.$$

Doksum (1967) shows that the asymptotic efficiency of T' relative to T is at least unity for all F and J.

Puri and Sen (1968) generalize Hollander's (1967) criteria to Chernoff-Savage type of scores. Let

$$Y_{i,u,v} = X_{iu}-X_{iv}, \quad u < v = 1,\cdots,J \quad \text{and} \quad i = 1,\cdots,I$$

and denote the c.d.f. of $Y_{i,u,v}$ by $G_{u,v}(x)$. Define the random variables

$$(1.5) \qquad S_{u,v} = (1/I) \sum_{i=1}^{I} E_{I,i} Z_{u,v,i}, \quad u < v = 1,\cdots,J.$$

where $Z_{u,v,i}$ is 1 or 0 according to whether the ith smallest order statistic in a sample of size I is from a positive or negative $Y_{i,u,v}$ and $E_{I,i}$ = expected value of ith smallest order statistic in a sample of size I drawn from the distribution

$$(1.6)\qquad \begin{aligned} \Psi^*(x) &= \Psi(x) - \Psi(-x), &\quad \text{for } x \geq 0 \\ &= 0 &, \quad \text{for } x < 0. \end{aligned}$$

Let

$$(1.7)\qquad V = \sum_{u<v} S_{u,v}.$$

If Ψ is uniform on $(-1,1)$ then V reduces to T given by (1.4). If Ψ is standard normal, V reduces to normal scores type of a statistic. Let

$$S_{u\cdot} = (1/J) \sum_{v=1}^{J} S_{u,v}, \qquad 1 \leq u \leq J$$

$$S^*_{u,v} = S_{u\cdot} - S_{v\cdot} \quad \text{for} \quad 1 \leq u \leq v \leq J,$$

and

$$(1.8)\qquad V^* = \sum_{u<v} S^*_{u,v}.$$

Notice that V and V^* are not distribution-free although they can be made asymptotically distribution-free by using consistent estimators for the variances of V and V^* under H_0. Puri and Sen (1968) show that the asymptotic efficiency of V^* relative to V is at least one for all G and all Ψ^{-1} satisfying certain regularity assumptions.

Shorack (1967) proposed a modified Friedman's chi-square test procedure using the 'amalgamation process' of Bartholomew (1959) [see also Barlow, et al. (1972), §1.2]. His procedure is as follows:

Replace each observation X_{ij} by r_{ij}, its rank in the ith row. Let

$$(1.9) \qquad \bar{r}_j = (1/I) \sum_{i=1}^{I} r_{ij}.$$

Now apply the amalgamation process to the $\bar{r}_j$'s and obtain m distinct integers $t_1, \cdots, t_m$ such that $t_1 + \cdots + t_m = J$ and m distinct quantities $\bar{r}_{[t_1]}, \cdots, \bar{r}_{[t_m]}$. Notice that the amalgamation process enables one to achieve the same order among the $\bar{r}_{[t_1]}, \cdots, \bar{r}_{[t_m]}$ as was postulated among the τ_j in H_1. Then reject H_0 for large values of the statistic

$$(1.10) \qquad \bar{\chi}^2 = \{12I/J(J+1)\} \sum_{j=1}^{m} t_j(\bar{r}_{[t_j]} - (J+1)/2)^2.$$

Shorack (1967) points out that it can be shown that the asymptotic efficiency of $\bar{\chi}^2$ relative to the corresponding parametric test (namely, his $\bar{B}$ test given on p. 1747, which is known as the $\bar{E}^2$ test) is

$$(1.11) \qquad \{J/(J+1)\} 12\sigma_F^2 [\int F'(t) dF(t)]^2$$

where σ_F^2 denotes the variance of F. Shorack (1967, pp. 1748 and 1742) also shows that under H_0 the distribution of $\bar{\chi}^2$ is a mixture of central chi-square distribution functions as $I \to \infty$.

So far we have been reviewing the results in the literature with

$n_{ij} = 1$ (that is, with one observation per cell). However, quite often the experimenter has more than one observation in each cell. Hettmansperger (1975) has extended Page's (1963) criterion to this situation. Let X_{ijk} be defined as in the beginning of the introduction. For each i, the X_{ijk} are ranked. Let $R_{ij\cdot}$ denote the sum of the ranks of the observations in the $(i,j)^{th}$ cell. Then $R_{ij\cdot}/n_{ij}$ is the average rank in the $(i,j)^{th}$ cell and $\sum_{i=1}^{I} (R_{ij\cdot}/n_{ij})$ is the total corresponding to treatment j. Hettmansperger (1985) proposes the test statistic

$$\tilde{T} = \sum_{j=1}^{J} j \sum_{i=1}^{I} (R_{ij\cdot}/n_{ij})$$

$$= \sum_{i=1}^{I} \sum_{j=1}^{J} (jR_{ij\cdot}/n_{ij}).$$

When the alternative hypothesis H_1 holds, we expect a high degree of agreement between the hypothesized ordering and the ordering exhibited by the treatment totals. Hence, we reject H_0 for large values of $\tilde{T}$. When $n_{ij} \equiv 1$, $\tilde{T}$ reduces to Page's (1963) statistic.

2. **RECENT WORK**. Mansouri-Ghiassi and Govindarajulu (1986) (to be abbreviated as MG-G (1986)) have proposed an asymptotically distribution-free test which is obtained by replacing the nuisance parameters by some consistent estimators and then employing a Kruskal-Wallis type of test procedure on the residuals. In other words, our approach is based on the "J-samplization" of the two-way layout. For a discussion of J-sample tests for ordered

alternatives, see, for instance, Govindarajulu and Haller (1977). Towards this, let $\{\hat{\beta}_i\}_{i=1}^{I}$ be some $n^{1/2}$-consistent estimators of $\{\beta_i\}_{i=1}^{I}$; i.e., given $\epsilon_i > 0$, there exists δ_i such that for n sufficiently large

$$(2.1) \qquad P(n^{1/2}|\hat{\beta}_i-\beta_i| \ge \delta_i) \le \epsilon_i, \quad i = 1,\cdots,I, \quad n = \Sigma\Sigma n_{ij}.$$

Let R_{ijk} = rank $(X_{ijk}-\hat{\beta}_i)$ in the overall ranking. Since rank $(X_{ijk}-\hat{\mu}-\hat{\beta}_i)$ = rank $(X_{ijk}-\hat{\beta}_i)$ throughout, we may without loss of generality assume that $\mu = 0$. Let

$$(2.2) \qquad \bar{R}_{\cdot j\cdot} = (1/(n_{\cdot j}))\sum_{i=1}^{I}\sum_{k=1}^{n_{ij}} R_{ijk} \quad (1 \le j \le J)$$

where $n_{\cdot j} = \sum_{i=1}^{I} n_{ij}$. That is, $\bar{R}_{\cdot j\cdot}$ denotes the average of ranks assigned to j^{th} treatment. We propose the test criterion

$$(2.3) \qquad L = \sum_{j=1}^{J} b_j\bar{R}_{\cdot j\cdot}$$

where $b_1 \le b_2 \le \cdots \le b_J$ are real-valued constants and not all b's are equal and we reject H_0 for large values of L. In particular, the $\hat{\beta}_i$ could be the least squares estimates of the block effects or the Lehmann (1963) robust estimates of the block effects. The latter are given by

$$\hat{\beta}_i = (IJ^2)^{-1} \sum_{j,i',j'} Y_{ij,i',j'}, \quad i = 1,2,\cdots,I$$

where

$$(2.4) \qquad Y_{ij,i',j'} = \text{median}\{X_{ijk}-X_{i'j'k'};\ 1 \le k,\ k' \le n_{ij},\ 1 \le i,\ i \le I,\ 1 \le j,\ j' \le J\}.$$

3. LIMITING DISTRIBUTION OF L. If $\beta_1, \beta_2, \cdots, \beta_I$ are known, let

$$(3.1) \qquad Q_{ijk} = \text{rank}\ (X_{ijk}-\beta_i),\quad 1 \le i \le I,\ 1 \le j \le J,\quad k = 1,\cdots,n_{ij}.$$

Define

$$(3.2) \qquad \underline{Q}' = (12/n)^{1/2}(\overline{Q}_{\cdot 1\cdot},\cdots,\overline{Q}_{\cdot J\cdot}),\quad \overline{Q}_{\cdot j\cdot} = (1/n_{\cdot j}) \sum_{i=1}^{I} \sum_{k=1}^{n_{ij}} Q_{ijk}.$$

It is well known that when $\lambda_{\cdot j} = \lim_{n\to\infty}(n_{\cdot j}/n)$ are bounded away from zero and one and F is continuous, $\underline{Q}$ is asymptotically distributed as multivariate normal with mean $E(\underline{Q})$ and a certain variance-covariance matrix. In particular, under H_0, the variance-covariance matrix is $\Sigma = ((\sigma_{j,j'}))$ where

$$(3.3) \qquad \sigma_{j,j'} = -1 + \delta_{jj'}(\lambda_{\cdot j}\lambda_{\cdot j'})^{-1/2}$$

and $\delta_{j,j'}$ is the Kronecker's delta function.

We have the following lemma towards the asymptotic behavior of $\underline{R} = (R_1,\cdots,R_J)$ where $R_j = (12/n)^{1/2}\ \overline{R}_{\cdot j\cdot}$ $(1 \le j \le J)$.

LEMMA 3.1. *Let* $\underline{R}$ *be as defined above. Let*

$$H_{ij,i',j'}(t_i-t_{i'}-\beta_i+\beta_{i'}) = E[\varphi(X_{ij1}-X_{i'j'1}-t_i+t_{i'})]$$

where the t_i's *are some real numbers and* $\varphi(u) = 1(0)$ *if* $u \geq 0$ $(u < 0)$.

Let $H = (H_1,\cdots,H_J)'$ *where*

$$(3.4)\quad H_j = (12/n)^{1/2}[(n_{\cdot j}+1)/2 + (1/n_{\cdot j}) \sum_{j'\neq j}^{J} \sum_{i,i'}^{I} n_{ij}n_{i'j'}H_{iji'j'}(\hat{\beta}_i-\hat{\beta}_{i'}-\beta_i+\beta_{i'})]$$

and $\hat{\beta}_i$ *and* $\hat{\beta}_{i'}$ *are some* $n^{1/2}$*-consistent estimates of* β_i *and* $\beta_{i'}$, *respectively. If* $\lambda_{ij} = \lim_{n\to\infty}(n_{ij}/n)$ *is bounded away from zero and one and* $\Sigma\lambda_{ij} = 1$, *and if the c.d.f.* $F(\cdot)$ *has a bounded density* $f(\cdot)$, *then*

$$(3.5)\qquad \underline{R}-\underline{H}-\underline{Q} + E(\underline{Q}) = o_p(1) \quad \text{as} \quad n \to \infty.$$

If, in addition, we have $n_{ij} = n_{i\cdot}n_{\cdot j}/n$ *and* f *is continuous, then*

$$(3.6)\qquad \underline{R}-\underline{Q} = o_p(1) \quad \text{as} \quad n \to \infty.$$

PROOF. See Lemmas 2.2 and 2.3 of MG-G (1986). Then we are led to the following theorem.

THEOREM 3.1. *If the assumptions of Lemma 3.1 are satisfied;*

(3.7) $$(\underline{b}' \Sigma \underline{b})^{-1/2}[L-(n/12)^{1/2}\underline{b}'E(\underline{Q})]$$

is asymptotically normal with mean $\underline{0}$ *and variance unity.*

PROOF. It follows from the asymptotic standard normality of

$$(\underline{b}' \Sigma \underline{b})^{-1/2}\underline{b}'[\underline{Q}-E(\underline{Q})]$$

and (3.6).

COROLLARY 3.1.1. *Since under* H_0,

$$\lim_{n\to\infty}[(1/n)\sum_{j=1}^{J}(1/n_{\cdot j})(nb_j-Jn_{\cdot j}\bar{b})^2 = \underline{b}' \Sigma \underline{b}$$

we have that

$$L^* = \left\{12/\left[\sum_{j=1}^{J}(1/n_{\cdot j})(nb_j-Jn_{\cdot j}\bar{b})^2\right]\right\}^{1/2}[L-(n/12)^{1/2}\underline{b}'E_0(\underline{Q})]$$

is asymptotically standard normal (see also Kruskal (1952)), where $J\bar{b} = \sum_1^J b_j$ *and* $E_0(Q_j) = (n+1)(3/n)^{1/2}$. *Hence we reject* H_0 *when* $L^* > z_\alpha$ *where* z_α *denotes the* $(1-\alpha)^{th}$ *quantile of the standard normal distribution and* α *denotes the level of significance.*

COROLLARY 3.1.2. *From Theorem 3.1 it follows that the test based on* L *is consistent.*

4. PITMAN EFFICACY OF L. Consider a sequence of "near" alternatives of the form

$$(4.1)\qquad H_n : \tau_{jn} = a_j n^{-1/2}\tau, \quad 1 \le j \le J.$$

Then the Pitman efficacy of L is given by

$$(4.2)\qquad \mathrm{eff}(L) = 12\left\{\left[\sum_{j=1}^{J} \lambda_{\cdot j}(a_j-\bar{a})(\lambda_{\cdot j}^{-1}b_j-J\bar{b})\right]^2 \Big/ \left[\sum_{j=1}^{J} \lambda_{\cdot j}(\lambda_{\cdot j}^{-1}b_j-J\bar{b})^2\right]\right\}\cdot [\textstyle\int f^2(x)dx]^2,$$

and it is maximized when

$$(4.3)\qquad b_j = \lambda_{\cdot j}(a_j-\bar{a}), \quad 1 \le j \le J$$

where $\bar{a} = \sum_{j=1}^{J} \lambda_{\cdot j}a_j$ and $\lambda_{\cdot j}$ and $\bar{b}$ are as defined earlier.

Without loss of generality, we can set $\bar{b} = 0$. With these optimal choices of b_j's we have

$$(4.4)\qquad \mathrm{eff}(L^*) = 12\left[\sum_{j=1}^{J} \lambda_{\cdot j}(a_j-\bar{a})^2\right][\textstyle\int f^2(x)dx]^2.$$

For the proofs of (4.2) and (4.4), the reader is referred to MG-G (1986, Theorem 3.1). Note that (4.4) with $a_j = j$ coincides with the efficacy of Hettmansperger's (1975) test when $n_{ij} = n/IJ$ and $n \to \infty$.

5. ASYMPTOTIC RELATIVE EFFICIENCY COMPARISONS. Assuming normality for F, MG-G (1986) derive the likelihood derivative test*(see Knoke (1975) for a reference on this) criterion t for testing $H_0 : \tau = 0$ and show its efficacy to be

$$\text{(5.1)} \qquad \text{eff}(t) = \sum_{j=1}^{J} \lambda_{\cdot j}(a_j - \bar{a})^2/\sigma^2.$$

It readily follows from (4.4) and (5.1) that the asymptotic efficiency of L relative to t is

$$\text{(5.2)} \qquad \text{ARE}(L, t) = 12\sigma^2\left[\int_{-\infty}^{\infty} f^2(x)dx\right]^2$$

which has been shown by Hodges and Lehmann (1956) to have a lower bound of 0.846.

In order to make asymptotic efficacy comparisons of L relative to the corresponding nonparametric tests for randomized blocks with one observation per cell, we assume that X_{ij} are independent with c.d.f. $P(X_{ij} \le x) = F(x-\beta_i-\tau_j)$ where $\tau_j = j\tau$ and we let I become large. Under this setup, our test statistic L becomes

$$L = \sum_{j=1}^{J} j\bar{R}_{\cdot j\cdot}.$$

Now by considering alternatives of the form $\tau_j = j\tau I^{-1/2}$ and

---

*The likelihood derivative test was proposed by C. R. Rao. The test criterion is the derivative of the likelihood function evaluated at the null hypothesis-value of the parameter of interest and the maximum likelihood estimates of the nuisance parameters, if any.

proceeding as in Section 4 with $I \to \infty$, it can be shown that

$$\text{eff}(L) = J(J+1)(J-1)[\int f^2(x)dx]. \tag{5.3}$$

Denoting the tests proposed by Jonckheere (1954), Page (1963), and Hollander (1967) by T, P and Y, respectively, we find that

$$\text{ARE}(L,T) = (2J+5)/2(J+1), \quad \text{ARE}(L,P) = (J+1)/J \tag{5.4}$$

and

$$\text{ARE}(L,Y) = [3+2(J-2)\rho^*(F)][f^2(x)dx]^2/\{2(J+1)[\int g^2(x)dx]\} \tag{5.5}$$

where $g(x) = dG(x)/dx$, and $G(x)$ is the c.d.f. of X_1-X_2, $\rho^*(F) = 12\eta(F)-3$ and

$$\eta(F) = P(X_1-X_2 < X_3-X_4 \text{ and } X_1-X_5 < X_6-X_7),$$

and where $X_1, X_2, \cdots, X_7$ are i.i.d. random variables with continuous c.d.f. $F(x)$. From (5.4) one can readily see that

$$1 \le \{\text{ARE}(L,T), \text{ARE}(L,P)\} \le 1.5.$$

The ARE(L,Y) is tabulated in Table 5.1, assuming some specific forms for F.

Table 5.1. ARE(L,Y).

F	$\eta(F)$	$\rho^*(F)$	ARE(L,Y)	Range of ARE(L,Y)
Uniform(-1,1)	.2909	.4909	[27+8.8362(J-2)]/8(J+1)	1.1045-1.125
Normal$(0,\sigma^2)$	.2902	.4824	[3+.9648(J-2)]/(J+1)	.9648-1.00
Negative exponential	.2894	.4728	[6+1.8912(J-2)]/(J+1)	1.8912-2.00
Double exponential	.2865	.4350	[7.68+2.2272(J-2)]/(J+1)	1.1136-1.28

From Table 5.1, we surmise that the test L compares quite favorably with Hollander's (1967) test. Perhaps this may be attributed to the fact that the ratios of the τ_j are known constants in our case.

REFERENCES

Barlow, R.E., Bartholomew, D.J., Bremner, J.M. & Brunk, H.D. (1972). *Statistical Inference Under Order Restrictions*. John Wiley & Sons, New York.

Bartholomew, D.J. (1959). A test of homogeneity for ordered alternatives. *Biometrika* **46**, 36-48.

Doksum, K.A. (1967). Robust procedures for some linear models with one observation per cell. *Ann. Math. Statist.* **38**, 878-883.

Govindarajulu, Z. & Haller, H.S. (1977). C-sample tests of homogeneity against ordered alternatives. *Proceedings of the Symposium to Honour Jerzy Neyman* (R. Bartoszynski, et al., eds.). Polish Scientific Publishers, Warszawa, 91-102.

Hettmansperger, T.P. (1975). Non-parametric inference for ordered alternatives in a randomized block design. *Psychometrika* **40**, 53-62.

Hodges, J.L. & Lehmann, E.L. (1956). The efficiency of some parametric competitors of the t-test. *Ann. Math. Statist.* **27**, 324-335.

Hollander, M. (1967). Rank tests for randomized blocks when the alternatives have an a priori ordering. *Ann. Math. Statist.* **38**, 867-877.

Jonckheere, A.R. (1954). A test of significance for the relation between m rankings and k ranked categories. *Brit. J. Statist. Psych.* **1**, 93-100.

Lehmann, E.L. (1963). Robust estimation in analysis of variance. *Ann. Math. Statist.* **34**, 957-966.

Knoke, J.D. (1975). Testing for randomness against autocorrelated alternatives: the parametric case. *Biometrika* **62**, 571-575.

Mansouri-Ghiassi, S.H. & Govindarajulu, Z. (1986). An asymptotically distribution-free test for ordered alternatives in two-way layouts. *J. Statist. Planning and Inference* **13**, 239-249.

Page, E.B. (1963). Ordered hypotheses for multiple treatments: a significance test for linear ranks. *J. Amer. Statist. Assoc.* **58**, 216-230.

Puri, M.L. & Sen, P.K. (1968). On Chernoff-Savage tests for ordered alternatives in randomized blocks. *Ann. Math. Statist.* **39**, 967-972.

Shorack, G.R. (1967). Testing against ordered alternatives in Model I analysis of variance: normal theory and nonparametric. *Ann. Math.* **38**, 1740-1752.

Z. Govindarajulu
Department of Statistics
University of Kentucky
Lexington, KY 40506-0027

S. H. Mansouri-Ghiassi
Department of Mathematics
Texas Tech. University
Lubbock, TX 79409

A CLASS OF DISTRIBUTION-FREE TESTS FOR TESTING HOMOGENEITY OF VARIANCES AGAINST ORDERED ALTERNATIVES

Subhash C. Kochar
Panjab University

R. P. Gupta
Dalhousie University

AMS 1980 subject classifications: *Primary* 62G10; *Secondary* 62G20.
Key words and phrases: *U-statistics, asymptotic relative efficiency.*

ABSTRACT

Let X_{ij}, $j = 1,2,\cdots,n_i$ be a random sample of size n_i from an absolutely continuous distribution with distribution-function $F(x/\sigma_i)$, with mean 0; $i = 1,2,\cdots,k$. The problem considered is to test the null hypothesis $H_0 : \sigma_1 = \sigma_2 = \cdots = \sigma_k$ against the alternative $H_1 : \sigma_1 \le \sigma_2 \le \cdots \le \sigma_k$. Let $2 \le c,d \le \min(n_1,n_2,\cdots,n_k)$ be two fixed integers and let $\phi(X_{i1},\cdots,X_{ic},X_{j1},\cdots,X_{jd})$ take value 1 (-1) when max as well as min of $(X_{i1},\cdots,X_{ic},X_{j1},\cdots,X_{jd})$ are both X_j's (X_i's) and it takes a value zero, otherwise. Let $U_{i,j}$ be the U-statistic corresponding to the kernel ϕ_{ij}. The proposed test statistic is $W_{c,d} = \sum_{i=1}^{k-1} a_i U_{i,i+1}$, with large values of $W_{c,d}$ leading to rejection of H_0. The optimum values of a_i's are obtained. The tests are quite efficient.

1. **INTRODUCTION.** Let $X_{i1},X_{i2},\cdots,X_{in_i}$ be independent random samples of size n_i from absolutely continuous distributions with distribution functions F_i, $i = 1,2,\cdots,k$. We assume that the F_i's

are identical in all aspects except for differing in dispersion, that is, $F_i(x) = F(x/\sigma_i)$, $i = 1,2,\cdots,k$. We shall focus on the case when the F_i's are symmetric, say around zero. We wish to test the null hypothesis

$$H_0: \sigma_1 = \sigma_2 = \cdots = \sigma_k \tag{1.1}$$

against the ordered alternative

$$H_1: \sigma_1 \le \sigma_2 \le \cdots \le \sigma_k \tag{1.2}$$

with at least one strict inequality.

This problem has been considered earlier by Govindarajulu and Haller (1977), Govindarajulu and Gupta (1978) and Rao (1982), among others. For testing H_0 against the ordered stochastic ordering alternative $H_2: F_1(x) \le F_2(x) \le \cdots \le F_k(x)$, tests have been proposed by Jonckheere (1954), Chacko (1963), Puri (1965) and Tryon and Hettmansperger (1973), among others.

In this paper, we propose a class of distribution-free tests based on weighted linear combinations of consecutive two-sample U-statistics for the above-mentioned testing problem. The tests are given in Section 2 and the distributions of the test statistics are discussed in Section 3. The optimal members in this class of tests are identified by obtaining the weighting coefficients which maximize the efficacy of the test in the proposed class when the scale parameters σ_i's are assumed to be equally spaced and the sample sizes are all equal. This is discussed in Section 4. In the last section, the proposed tests are compared with other known tests in the Pitman asymptotic relative efficiency sense and these are seen to be quite efficient when the underlying distributions are symmetric. It is observed that tests with large values of $c+d$ should be used for lighter-tailed distributions.

2. **THE PROPOSED TESTS.** Under the alternative H_1, for $i < j$, the X_j's are more dispersed than the X_i's. First we consider a two sample U-statistic which will test for differences in the dispersions of two populations and then treat the k-sample problem.

Let c and d be two fixed integers such that $2 \le c,d \le \min(n_1,\cdots,n_k)$. Define for $i < j$, $i,j = 1,2,\cdots,k$,

$$(2.1)\qquad \phi_{i,j}^{(c,d)}(X_{i\alpha_1},\cdots,X_{i\alpha_c};\ X_{j\beta_1},\cdots,X_{j\beta_d})$$

$$= \begin{cases} 1, & \text{if min and max}(X_{i\alpha_1},\cdots,X_{i\alpha_c};\ X_{j\beta_1},\cdots,X_{j\beta_d}) \\ & \text{are both } X_j\text{'s;} \\ -1, & \text{if min and max}(X_{i\alpha_1},\cdots,X_{i\alpha_c};\ X_{j\beta_1},\cdots,X_{j\beta_d}) \\ & \text{are both } X_i\text{'s;} \\ 0, & \text{otherwise.} \end{cases}$$

The two-sample U-statistic corresponding to the kernel $\phi_{ij}^{(c,d)}$ is given by

$$(2.2)\qquad U_{i,j}^{(c,d)} = \left[\binom{n_i}{c}\binom{n_j}{d}\right]^{-1} \sum_{c} \phi_{ij}^{(c,d)}(X_{i\alpha_1},\cdots,X_{i\alpha_c};\ X_{j\beta_1},\cdots,X_{j\beta_d})$$

where $\sum_{c}$ denotes summation extended over all combinations of c integers $(\alpha_1,\cdots,\alpha_c)$ chosen from $(1,2,\cdots,n_i)$ and all combinations of d integers $(\beta_1,\cdots,\beta_d)$ chosen from $(1,\cdots,n_j)$.

For $i < j$, X_j's are more dispersed than X_i's and this will be indicated by large values of $U_{i,j}^{(c,d)}$. Tests based on these statistics have been earlier proposed by Kochar and Gupta (1986) for the two-sample problem.

$U_{i,j}^{(c,d)}$ can also be expressed in terms of the ranks of the observations. Let $X_{i(\ell)}$ be the ℓth order statistic of the X_i sample and let $R_{i(\ell)}$ be its rank in the combined increasing arrangement of X_i's and X_j's. Similarly define $X_{j(m)}$ and $R_{j(m)}$, then

$$(2.3)\quad U_{i,j}^{(c,d)} = \left[\binom{n_i}{c}\binom{n_j}{d}\right]^{-1}\left[\sum_{m=1}^{n_j}\binom{m-1}{d-1}\binom{R_{j(m)}-m}{c} - \sum_{\ell=1}^{n_i}\binom{n_i-\ell}{c-1}\binom{n_j-R_{i(\ell)}+\ell}{d}\right].$$

For testing H_0 against H_1, we propose the class of statistics

$$(2.4)\qquad W_{c,d} = \sum_{i=1}^{k-1} a_i U_{i,i+1}$$

where $a_1,\cdots,a_{k-1}$ are suitably chosen real constants. For the convenience of notation we shall be omitting the superscript (c,d) in $U_{i,j}^{(c,d)}$ and $\phi_{ij}^{(c,d)}$. For each vector $(a_i,\cdots,a_{k-1})$ and for fixed c and d, we get a distinct member of this class of test statistics. Large values of $W_{c,d}$ lead to rejection of H_0.

3. **DISTRIBUTION OF $W_{c,d}$.** Clearly,

$$(3.1)\qquad E[W_{c,d}] = \sum_{i=1}^{k-1} a_i \mu_{i,i+1}$$

where

(3.1)
$$\mu_{i,i+1} = E[U_{i,i+1}]$$
$$= d\Big[\int_{-\infty}^{\infty} F_i^c(x)\, F_{i+1}^{d-1}(x)\, dF_{i+1}(x)$$
$$-\int_{-\infty}^{\infty}\{1-\overline{F}_i^c(x)\}\, \overline{F}_{i+1}^{d-1}(x)dF_{i+1}(x)\},$$

and

$$\overline{F} = 1-F.$$

Under H_0,

(3.3)
$$E[W_{c,d}] = \frac{d-c}{d+c}\sum_{i=1}^{k-1} a_i.$$

Let

(3.4)
$$\underline{U}' = (U_{1,2};\ U_{2,3},\cdots,U_{k-1,k}).$$

Since the $U_{i,j}$'s are two-sample U-statistics, the joint limiting normality of $\{U_{ij}\}$ follows immediately (see Lehmann (1963)) as stated in the theorem below.

THEOREM 3.1. *The asymptotic distribution of* $\sqrt{N}[\underline{U}-E(\underline{U})]$ *as* $N \longrightarrow \infty$ *in such a way that* $n_i/N \longrightarrow p_i$, $0 < p_i < 1$ *for* $i = 1,\cdots,k$, *is multivariate normal with mean vector* $\underline{0}$ *and dispersion matrix* $\Sigma = ((\sigma_{ij}))$, *where* $N = \sum_{i=1}^{k} n_i$ *and*

(3.5)
$$\sigma_{ij} = \begin{cases} \dfrac{c^2}{p_i}\,\xi^{(i)}_{i,i+1;i,i+1} + \dfrac{d^2}{p_{i+1}}\,\xi^{(i+1)}_{i,i+1;i,i+1} & \text{for } i = j = 1,2,\cdots,k-1 \\ \dfrac{cd}{p_{i+1}}\,\xi^{(i+1)}_{i,i+1;i+1,i+2} & \text{for } j = i+1,\ i = 1,2,\cdots,k-2, \\ \dfrac{cd}{p_i}\,\xi^{(i)}_{i-1,i;i,i+1} & \text{for } j = i-1,\ i = 2,3,\cdots,k-1, \\ 0 & \text{otherwise} \end{cases}$$

where

$$\xi^{(i)}_{i,i+1;i,i+1} = E[\{\psi^{(i)}_{i,i+1}(X)\}^2] - E^2[U_{i,i+1}], \tag{3.6}$$

$$\xi^{(i+1)}_{i,i+1;i,i+1} = E[\{\psi^{(i+1)}_{i,i+1}(X)\}^2] - E^2[U_{i,i+1}] \tag{3.7}$$

$$\xi^{(i+1)}_{i,i+1;i+1,i+2} = E[\psi^{(i+1)}_{i,i+1}(X)\ \psi^{(i+1)}_{i+1,i+2}(X)] - E[U_{i,i+1}]E[U_{i+1,i+2}], \tag{3.8}$$

$$\psi^{(i)}_{i,j}(x) = E[\phi_{i,j}(x, X_{i2}, \cdots, X_{ic};\ X_{j1}, \cdots, X_{jd}], \tag{3.9}$$

$$\psi^{(j)}_{i,j}(x) = E[\phi_{i,j}(X_{i1}, \cdots, X_{ic};\ x, X_{j2}, \cdots, X_{jd})]. \tag{3.10}$$

It can be seen that under H_0,

$$\sigma_{ij} = \begin{cases} 2c^2d^2\rho_m\left\{\dfrac{1}{p_i} + \dfrac{1}{p_{i+1}}\right\}, & \text{for } i = j = 1,2,\cdots,k-1, \\ \dfrac{-2c^2d^2\rho_m}{p_{i+1}} & \text{for } j = i+1,\ i = 1,2,\cdots,k-2, \\ \dfrac{-2c^2d^2}{p_i}\rho_m & \text{for } j = i-1,\ i = 2,3,\cdots,k-1, \\ 0 & \text{otherwise;} \end{cases} \tag{3.11}$$

where

$$m = c+d$$

and

$$\rho_m = \frac{1}{(m-1)^2}\left[\frac{1}{2m-1} - \frac{2}{m^2} + \frac{((m-1)!)^2}{(2m-1)!}\right]. \tag{3.12}$$

In case all the sample sizes are equal, that is, $p_1 = p_2 = \cdots = p_k = \frac{1}{k}$, (3.11) becomes

$$(3.13) \qquad \sigma_{ij} = \begin{cases} 4c^2d^2k\rho_m, & \text{for } i = j = 1,2,\cdots,k-1, \\ -2c^2d^2k\rho_m, & \text{for } j = i+1,\ i = 1,2,\cdots,k-2, \\ -2c^2d^2k\rho_m, & \text{for } j = i-1,\ i = 1,2,\cdots,k-1, \\ 0 \quad , & \text{otherwise.} \end{cases}$$

Since $W_{c,d}$ is a linear combination of the components of $\underline{U}$, the proof of the following theorem follows from the transformation theorem (Serfling (1980), p. 122).

THEOREM 3.2. *The asymptotic distribution of* $N^{1/2}[W_{c,d}-E(W_{c,d})]$ *as* $N \longrightarrow \infty$ *in such a way that* $n_i/N \longrightarrow p_i$, $0 < p_i < 1$, $i = 1,2,\cdots,k$, *is normal with mean zero and variance* $\underline{a}'\Sigma\underline{a}$. *Under* H_0, $E[W_{c,d}] = \frac{d-c}{d+c} \sum_{i=1}^{k-1} a_i$ *and* $\underline{a}'\Sigma\underline{a} = 4c^2d^2k\rho_m\left[\sum_{i=1}^{k-1} a_i^2 - \sum_{i=1}^{k-2} a_ia_{i+1}\right]$, *when* $p_i = 1/k$, $i = 1,2,\cdots,k$.

4. OPTIMAL CHOICE OF WEIGHTS. Now we consider the problem of obtaining the optimal weights, a_i, so that for fixed c and d the test $W_{c,d}$ has maximum efficacy for the sequence of Pitman type alternatives

$$(4.1) \qquad H_N\colon\ F_i(x) = F[x/(\sigma+N^{-1/2}\delta_i)],\quad i = 1,\cdots,k$$

where δ_i and σ are real positive constants. We assume without loss of generality that $\sigma = 1$, since all relative orderings, and hence the $W_{c,d}$, remains invariant if the variables are all multiplied by the same positive constant. Further, for efficiency comparisons, we consider the equal sample size case, that is, $p_i = 1/k$, $i = 1,\cdots,k$ and the equally spaced alternatives of the type $\delta_i = i\delta$, $\delta > 0$, for $i = 1,\cdots,k$. Thus the alternative H_N becomes

$$(4.2) \qquad H_N' : F_i(x) = F[x/(1+N^{-1/2}i\delta)].$$

The following theorem gives the asymptotic distribution of $\underline{U}$ under the sequence of alternatives $\{H_N'\}$. The proof is routine and hence is omitted.

THEOREM 4.1. *Let* X_{ij} *be independent random variables with cumulative distribution function* $F_i(x)$, $j = 1,\cdots,n_i$, $i = 1,\cdots,k$, *where* F_i's *are given by* (4.2). *Under the following assumptions the limiting distribution of* $N^{1/2}[\underline{U} - \frac{d-c}{d+c} J_{k-1}]$ *is* $(k-1)$*-dimensional multivariate normal with mean vector* μJ_{k-1} *and dispersion matrix* $\Sigma = ((\sigma_{ij}))$ *given by* (3.13):

(i) F *is absolutely continuous with density* f;

(ii) $\left|\frac{f(x)-f(x+h)}{h}\right| \le g(x)$ *for small* h *and*

$$\int_{-\infty}^{\infty} x[g(x)]^i f(x)dx < \infty \quad \textit{for} \quad i = 1,\cdots,m.$$

Here $J_{k-1} = [1]_{(k-1)\times 1}$.

$$(4.3) \qquad \mu = cd\delta \int_{-\infty}^{\infty} xf(x) \{F^{m-2}(x) - \bar{F}^{m-2}(x)\}dF(x)$$

and $m = c+d$.

In the following theorem we obtain the optimum weights a_i's of the test statistics $W_{c,d}$ so that they have the maximum efficacy.

THEOREM 4.2. *Under the assumptions of Theorem* 4.1 *and under the sequence of alternatives* $\{H'_N\}$ *given by* (4.2), *the efficacy of the test* $W_{c,d}$ *is maximized when*

$$a_i = i(k-i)/2k, \quad i = 1,\cdots,k-1. \tag{4.4}$$

PROOF. Since $W_{c,d}$ is a linear combination of components of $\underline{U}$, it follows from the above theorem that $N^{1/2}\left[W_{c,d} - \frac{d-c}{d+c}\sum_{i=1}^{k-1} a_i\right]$ is asumptotically normally distributed with mean $\mu\left[\sum_{i=1}^{k-1} a_i\right]$ and variance $\underline{a}'\Sigma\underline{a}$.

Let $\theta = N^{-1/2}\delta$. Then the efficacy of $W_{c,d}$ (see Govindarajulu and Gupta (1978) and Rao (1982)) is given by

$$e(W_{c,d}) = \left[\frac{d}{d\theta} E_{H'_N}(W_{c,d})\Big|_{\theta=0}\right]^2 \Big/ \underline{a}'\Sigma\underline{a} \tag{4.5}$$

$$= \frac{c^2 d^2\left[\sum_{i=1}^{k-1} a_i\right]^2 G_m^2}{\underline{a}'\Sigma\underline{a}}$$

$$= \frac{G_m^2\left[\sum_{i=1}^{k-1} a_i\right]^2}{2\rho_m \underline{a}'\Sigma^{*}\underline{a}}$$

where

$$\Sigma^* = ((\sigma^*_{ij}))$$

and

$$(4.6) \qquad \sigma^*_{ij} = \begin{cases} 2k, & \text{if } j = i,\ i = 1,\cdots,k-1 \\ -k, & \text{if } j = i+1,\ i = 1,\cdots,k-2, \\ -k, & \text{if } j = i-1,\ i = 2,\cdots,k-1, \\ 0, & \text{otherwise,} \end{cases}$$

and

$$(4.7) \qquad G_m^2 = \int_{-\infty}^{\infty} xf^2(x)\ \{F^{m-2}(x) - \bar{F}^{m-2}(x)\}dx.$$

It is seen from (4.5) that the efficacy of $W_{c,d}$ depends upon c and d only through the sum $m = c+d$.

$e(W_{c,d})$ is maximized when $\left[\sum_1^{k-1} a_i\right]^2 / \underline{a}'\Sigma^*\underline{a}$ is maximized with respect to $\underline{a}$. This is maximized when

$$(4.8) \qquad \underline{a} = \Sigma^{*-1}\ J_{k-1}$$

(see Rao (1973), p. 60) and

$$(4.9) \qquad \sup_{\underline{a}} e(W_{c,d}) = \frac{G_m^2}{2\rho_m} J'_{k-1}\ \Sigma^{*-1}\ J_{k-1}.$$

Also it is known that (see Graybill (1969))

$$\Sigma^{*-1} = ((\sigma^{*ij}))$$

where

$$(4.10) \qquad \sigma^{*ij} = \begin{cases} i(k-j)/k^2 & \text{if } i \le j \\ j(k-i)/k^2 & \text{if } i \ge j \end{cases}.$$

It follows from (4.8) to (4.10) that the optimum choice of a_i's

$$a_i^* = i(k-i)/2k, \quad i = 1,\cdots,k-1 \tag{4.11}$$

and

$$J'_{k-1}\ \Sigma^{*-1}\ J_{k-1} = (k^2-1)/12$$

and the efficacy of the optimum test $W^*_{c,d}$ with this weighting coefficient is

$$e(W^*_{c,d}) = \frac{(k^2-1)G_m^2}{24\rho_m}, \tag{4.12}$$

which depends upon c and d only through the sum $m = (c+d)$. For some standard distributions, we computed the values of $e(W^*_{c,d})$ and obtained the optimum values of m for which $e(W^*_{c,d})$ is maximized. These optimum values m^* of m are given below in Table 1 and Table 2.

It is seen from these tables that for thin-tailed distributions $W^*_{c,d}$ tests with large values of m are more appropriate.

5. **ASYMPTOTIC RELATIVE EFFICIENCIES**. In this section, we compare the asymptotic efficiencies of the $W^*_{c,d}$ tests relative to some other known nonparametric tests for this testing problem.

Govindarajulu and Gupta (1978) developed a locally most powerful rank test S_{1N} for normal distributions and a class of weighted sum of Chernoff-Savage type statistics S_{3N} for testing H_0 against H_1. It can be seen that $e(W^*_{2,2}) = 15(k^2-1)\left\{\int_{-\infty}^{-\infty} xf^2(x)[2F(x)-1]\right\}$ and $W^*_{2,2}$ is as efficient as the statistic S_{3N}, specialized to Mood type scores as considered by Govindarajulu and Gupta (1978).

Table 1 gives the asymptotic efficiencies of $W^*_{c,d}$ tests relative to the S_{1N} test for some standard symmetric distributions.

Table 1. Asymptotic relative efficiencies of $W^*_{c,d}$ relative to S_{1N}.

Distribution	m^*	2	3	4	5	6
Double exponential	11	.9161	.9776	.9735	.9704	.9692
Normal	16	.9311	.9310	.9313	.9307	.9308

Distribution	m^*	7	8	9	10
Double exponential	11	.9690	.9687	.9883	.9171
Normal	16	.9309	.9310	.9306	.9312

Rao (1982) proposed the K_2-test based on linear combinations of the generalized forms of Sugiura (1965) type U-statistics for this problem. The efficacy of Rao's test is

$$e(K_2) = \frac{(k^2-1)(2k-1)^2 G_{2k}^2}{12A(2k-1,2k-1)} \tag{5.1}$$

where

$$A(2k-1,2k-1) = \frac{(2k-1)^2}{2k^2(4k-1)} - \frac{1}{2k^2} + \frac{2[(2k-1)!]^2}{(4k-1)!}. \tag{5.2}$$

From (4.12) and (5.1), we find that the asymptotic relative efficiency of $W^*_{c,d}$ relative to K_2 is

(5.3) $$e(W^*_{c,d}, K_2) = \frac{A(2k-1,2k-1)G_m^2}{2(2k-1)^2 \rho_m G_{2k}^2}.$$

The ARE's of the $W^*_{c,d}$ test with respect to Rao's K_2 test for some standard distributions are given in Table 2.

Table 2. Asymptotic relative efficiency of $W^*_{c,d}$ relative to Rao's test.

Distribution	m^*	2	3	4	5	6
Logistic	11	1.0990	1.0510	1.0149	1.0010	1.0019
Double exponential	11	1.0190	1.0456	1.0130	1.0004	1.0023
Normal	16	1.2251	1.1466	1.0780	1.0364	1.0139
Distribution	m^*	7	8	9	10	
Logistic	11	1.0119	1.0274	1.0462	1.0673	
Double exponential	11	1.0125	1.0273	1.0445	1.0640	
Normal	16	1.0032	1.0000	1.0017	1.0067	

It is clear from these tables that the newly proposed tests $W^*_{c,d}$ are quite efficient for testing H_0 against H_1 for symmetric distributions. For lighter tailed distributions $W^*_{c,d}$ statistics with comparatively large values of m should be used. The performance of these tests for nonsymmetric distributions is similar to what is observed in Kochar and Gupta (1986). These tests are not appropriate for such distributions. In fact, tests based on kernels which look at only the subsample maxima should be used (see Kochar (1978)).

This work was carried out when the first author was a visiting fellow at Dalhousie University and was supported by NSERC grant no.

A5290. The authors are grateful to the referee for making valuable suggestions to an earlier draft of the paper.

REFERENCES

Chacko, V.J. (1963). Testing homogeneity against ordered alternatives. *Ann. Math. Statist.* **34**, 945-956.

Govindarajulu, Z. & Gupta, G.D. (1978). Tests for homogeneity of scale against ordered alternatives. *Trans. 8th Prague Conference on Information Theory, Statistical Decision Functions, Random Processes* (eds. J. Kozesnik, et al.). Academia Publ. House: Prague, Vol. **A**, 235-245.

Govindarajulu, Z. & Haller, H.S. (1977). c-sample tests of homogeneity against ordered alternatives. *Proceedings of the Symposium to Honour Jerzy Neymann* (eds. R. Bartoszynski, et al.). Polish Scientific Publ.: Warszawa, 91-102.

Graybill, F.A. (1969). *Introduction to Matrices with Applications in Statistics*. Wordsworth Publ. Co.: California.

Jonckheere, A.R. (1954). A distribution-free k samples test against ordered alternatives. *Biometrika* **41**, 133-145.

Kochar, S.C. (1978). A class of distribution-free tests for the two-sample slippage problem. *Commun. Statist. Theor.-Meth.* **A(7)**, 1234-1252.

Kochar, S.C. & Gupta, R.P. (1986). Some competitors of the Mood test for the two-sample scale problem. *Commun. Statist. Theor.-Meth.* **15** (1), 231-239.

Lehmann, E.L. (1963). Robust estimation in analysis of variance. *Ann. Math. Statist.* **34**, 957-966.

Puri, M.L. (1965). Some distribution-free k sample rank order tests of homogeneity against ordered alternatives. *Comm. Pure Appl. Math.* **18**, 51-63.

Rao, C.R. (1973). *Linear Statistical Inference and its Applications*. John Wiley: New York.

Rao, K.S.M. (1982). Nonparametric tests for homogeneity of scale against ordered alternatives. *Ann. Inst. Statist. Math. Part* **A.34**, 51-58.

Serfling, R.J. (1980). *Approximation Theorems of Mathematical Statistics*. John Wiley: New York.

Suguira, N. (1965). Multisample and multivariate nonparametric tests based on U-statistics and their asymptotic efficiencies. *Osaka J. Math.* **2**, 385-426.

Tryon, P.V. & Hettmansperger, T.P. (1973). A class of nonparametric tests for homogeneity against ordered alternative. *Ann. Statist.* 1, 1061-1070.

Subhash C. Kochar
Dept of Statistics
Panjab University
Chandigarh-160014
India

R.P. Gupta
Dept. of Mathematics, Statistics and Computing Science
Dalhousie University
Halifax, N.S. Canada B3H 3J5

ROBUSTNESS OF $\bar{\chi}^2$ AND $\bar{E}^2$: A CASE FOR A RANKED TEST IN ORDER-RESTRICTED INFERENCE

Michael J. Schell
St. Jude Children's Research Hospital

Bahadur Singh
Memphis State University

AMS 1980 subject classification: Primary 62G10; Secondary 65C05.
Keywords and phrases: chi-bar square distribution, chi-bar-square rank distribution, E-bar-square distribution, order restricted inference, power function, robustness.

ABSTRACT

In order restricted inference, the $\bar{\chi}^2$ and $\bar{E}^2$ test statistics, which assume an underlying normal distribution, are commonly used to test the null hypothesis of equality of means for k treatments against some given ordered alternative. We note the following results from Monte Carlo analysis. The $\bar{\chi}^2$ test is anticonservative for highly leptokurtic distributions (kurtosis ≥ 6), while the $\bar{E}$ test is conservative. A nonparametric test, $\bar{\chi}^2_{\text{rank}}$, avoids these robustness concerns and provides larger power for mild alternatives than either $\bar{\chi}^2$ or $\bar{E}^2$ when the distributions are highly leptokurtic.

1. **INTRODUCTION.** In order restricted inference, the two most widely used statistics for testing the equality of means from k populations against some given ordered alternative are $\bar{\chi}^2$ and $\bar{E}^2$, which assume that the treatment variances are known and unknown, respectively. Both tests also assume that the treatment observations have a normal distribution, as discussed at length by Barlow et al.

(1972). Many tests based on the normal distribution are not robust with respect to departures from normality, particularly when the distributions are heavy-tailed. A nonparametric alternative, in which raw values are replaced by ranks in the $\bar{\chi}^2$ statistic, has largely been ignored in the literature. In this paper we compare the robustness and power of the three tests, under different distributional assumptions, using Monte Carlo techniques.

2. THREE ORDER RESTRICTED TEST STATISTICS. Let X_{ij}, $i = 1, \cdots, k$; $j = 1, \cdots, n_i$ be k independent random samples of size n_i from the ith treatment with some continuous underlying distribution function F, mean μ_i and common variance σ^2. In this paper we concern ourselves only with the simple order alternative, that is, testing

$$H_0 : \mu_1 = \mu_2 = \cdots = \mu_k$$

versus

$$H_1 : \mu_1 \le \mu_2 \le \cdots \le \mu_k,$$

with at least one strict inequality.

Let $\bar{\mu} = (\bar{\mu}_1, \cdots, \bar{\mu}_k)$ be the maximum likelihood estimate of $\mu = (\mu_1, \cdots, \mu_k)$, subject to the restriction $H_0 \cup H_1$, and let $P(\ell, k; w)$ denote the probability, under H_0, that $\bar{\mu}$ has exactly ℓ distinct values. The likelihood ratio tests reject H_0 for large values of

$$\bar{\chi}^2 = \sum_{i=1}^{k} w_i (\bar{\mu}_i - \hat{\mu})^2, \tag{1}$$

where $w_i = n_i/\sigma^2$ and $\hat{\mu} = \sum_{i=1}^{k} \sum_{j=1}^{n_i} w_i X_{ij} / \sum_{i=1}^{k} w_i$, when σ^2 is known and

$$(2) \qquad \bar{E}^2 = \sum_{i=1}^{k} w_i(\bar{\mu}_i - \hat{\mu})^2 \Big/ \sum_{i=1}^{k} \sum_{j=1}^{n_i} (X_{ij} - \hat{\mu})^2,$$

where $w_i = n_i$, when σ^2 is unknown. When F is normal,

$$(3) \qquad \Pr(\bar{\chi}^2 \geq x) = \sum_{\ell=2}^{k} P(\ell, k; w) \Pr(\chi^2_{\ell-1} \geq x), \quad x > 0,$$

$$\Pr(\bar{\chi}^2 = 0) = P(1, k; w),$$

where $\chi^2_{\ell-1}$ is a chi-square random variable having $\ell-1$ degrees of freedom and

$$(4) \qquad \Pr(\bar{E}^2 \geq x) = P(\ell, k; w) \Pr(B_{1/2(\ell-1), 1/2(N-\ell)} \geq x), \ x > 0,$$

$$\Pr(\bar{E}^2 = 0) = P(1, k; w),$$

where $B_{a,b}$ denotes a random variable having a beta distribution with parameters a and b and $N = \sum_{i=1}^{k} n_i$.

The distribution-free alternative to $\bar{\chi}^2$, denoted $\bar{\chi}^2_{rank}$, in which ranks replace the raw data, was introduced by Chacko (1963). Early work on the properties of $\bar{\chi}^2_{rank}$ is well summarized by Barlow et al. (1972, Section 4.4). Assume the same setting as presented above. Then let R_{ij} denote the rank (from lowest to highest) of

X_{ij} among all the observations. Letting $N = \sum_{i=1}^{k} n_i$ and treating R_{ij} as a random variable, R_{ij} has mean $(N+1)/2$ and variance $\sigma^2_{R_{ij}} = (N+1)(N-1)/12$ under H_0. However, to be consistent with the existing definition of $\bar{\chi}^2_{\text{rank}}$ and to obtain a distribution closer to the distribution for small n_i, the weights will be taken to be $w_i = n_i/((N/(N-1))\hat{\sigma}^2_{R_{ij}}) = 12n_i/(N(N+1))$. Thus

$$\bar{\chi}^2_{\text{rank}} = \frac{12}{N(N+1)} \sum_{i=1}^{k} n_i \left[\bar{\mu}_i - \frac{N+1}{2}\right]^2,$$

where $\bar{\mu}_i$ represents the amalgamated means based on the ranks. It has been shown that the limiting distribution of $\bar{\chi}^2_{\text{rank}}$ is $\bar{\chi}^2$, where $N \longrightarrow \infty$, with $\lim_{N\to\infty} n_i/N = \nu_i$. For small sample sizes, however, the distributions can be quite different, as noted by Parsons (1979, 1981). The number of distinct rankings is given by $N!/\prod_{i=1}^{k} n_i!$. However, many different rankings yield the same value for $\bar{\chi}^2_{\text{rank}}$. For example, when $k = 3$ and $n = 5$, 756,756 distinct rankings give rise to 177 different $\bar{\chi}^2_{\text{rank}}$ values. When $k = 5$ and $n = 5$, 6.2×10^{14} distinct rankings give rise to about 2500 $\bar{\chi}^2_{\text{rank}}$ values.

Although the $\bar{\chi}^2$ and $\bar{E}^2$ distributions were derived by assuming F to be normal, it is sufficient that the sample means be normal. Hence, by the central limit theorem, the distributions given above are the limiting distributions as $n_i \longrightarrow \infty$ for fixed number treatments, k, when F is nonnormal. It is also known (Barlow et al. (1972), p. 191), for $n_i = 1$, $i = 1,\cdots,k$ and F, an exponential family distribution, that

$$\lim_{k\to\infty} \Pr\{-2 \ln \lambda \le c \mid \ell = \ell_0\} = \Pr\{\chi^2_{\ell_0} \le c\},$$

where λ is the likelihood ratio statistic and c is a constant for $\ell_0 = 1,2,\cdots$. Furthermore, given a simple order alternative and equal weights, the $P(\ell,k)$'s are distribution-free. From these results and the robustness of the χ^2 and F tests ($\bar{E}^2$ is functionally related to the F test), Barlow et al. (1972) concluded that "$\bar{\chi}^2$ and $\bar{E}^2$ are likely to be robust, particularly when the weights are equal and k is large."

It is more common in practice for the number of treatments to be small, especially in medical applications where the treatment modifications might be different doses of a drug. In this paper, the case of three treatments is studied and equal weights have been assumed (hence, let $n_i = n$, $i = 1,\cdots,k$).

3. SIMULATION METHODS AND RESULTS. In this section we present the results obtained with the use of Monte Carlo methods.

3.1 ***Methods***. To assess the robustness of the $\bar{\chi}^2$ and $\bar{E}^2$ tests with respect to departures from normality and to compare the power functions to each other and $\bar{\chi}^2_{rank}$, we undertook a Monte Carlo study. Eight nonnormal underlying distributions were considered: the uniform, the exponential, the double exponential, the lognormal, the t-distribution with 3 degrees of freedom (t_3) and three normal distributions, described below. Let

$$Y = pX_1 + qX_2 \quad \text{where} \quad X_1 \sim N(\delta q, 1) \quad \text{and} \quad X_2 \sim (-\delta p, r^2)$$

denote a normal mixture, where $q = 1-p$. Then Y is characterized by three parameters: p, r and δ. In this study, $\delta = 1$ was fixed. Thus the distributions will be referred to as NM(p,r). The three choices used were NM(.1,4), NM(.9,4) and NM(.5,2). Since kurtosis will be seen to play an important role in the simulation study, the parameters of the lognormal distribution were chosen to have kurtosis equal to that of the double exponential. Thus, the density function for the lognormal chosen is $f(x) = (x\sigma\sqrt{2\pi})^{-1} \exp(-(\ln(x)-\eta)^2/2\sigma^2)$, $x > 0$, where $\eta = 0$ and $\sigma^2 = 0.150$. Table 3 gives the skewness and kurtosis for each distribution mentioned.

The distributions of $\bar{\chi}^2$ and $\bar{E}^2$ under H_0 when F is normal are derived from equations (3) and (4), respectively. The exact distribution of $\bar{\chi}^2_{rank}$ can be obtained by the method of Parsons (1979). The exact distribution of $\bar{\chi}^2_{rank}$ is discrete and does not lead to exact α-level tests. Thus, to improve the comparison, we employed a randomized test which yields exact α-level tests when $n = 5$. For larger n, the $\bar{\chi}^2$ approximation was used since the exact distribution is cumbersome to obtain.

As an indication of robustness in the tail, we computed

$$z = (p-\pi)/\sqrt{\pi(1-\pi)/N},$$

where p is the simulated power, π is the theoretical power of the $\bar{\chi}^2$ statistic and N is the number of repetitions used. When p is a sample proportion with mean π, a z value exceeding 2.58 in absolute value is generated with probability .01. We call such an event a departure from robustness. We also define p/π to be the relative power for the simulation.

To assess the convergence over the entire distribution to the $\bar{\chi}^2$ distribution, we partitioned simulated $\bar{\chi}^2$ values into the following intervals: exactly zero, $0^+ - 0.5$, $0.5^+ - 1.0, \cdots, 9.5^+ - 10.0$, and greater than 10.0. The probability that the $\bar{\chi}^2$ statistic is in each class was computed from equation (1). Let $F(x)$ denote the distribution function of $\bar{\chi}^2$, $f(x) = (F(x)-F(x-h))/h$ be an approximation to the density function and $H(x) = N(F(x)-F(x-h))$, for N realizations of $\bar{\chi}^2$. Let $\hat{F}(x)$, $\hat{f}(x)$ and $\hat{H}(x)$ denote the simulated distribution function, density function and observed number of values in $(x-h, x]$, given N realizations of the statistic being studied, respectively. Finally, let

$$KS = \max_m |F(m/2)-\hat{F}(m/2)|, \text{ a Kolmogorov-Smirnov type statistic,}$$

$$L_1 = \sum_{m=0}^{21} |f(m/2)-\hat{f}(m/2)|,$$

$$L_2 = \sum_{m=0}^{21} (f(m/2)-\hat{f}(m/2))^2/h,$$

and

$$\chi^2 = \sum_{m=0}^{21} (H(m/2)-\hat{H}(m/2))^2/H(m/2),$$

where $h = 1/2$ and $F(21/2) = \hat{F}(21/2)$ are redefined to be 1. The critical values were chosen to be 0.350 for L_1, 0.250 for L_2, 0.0129 for KS and 38.9 for χ^2. The first two levels were chosen in an ad hoc fashion, based on the variability seen in the statistic when F is normal. The latter two critical levels were chosen so that, when F is normal, the probability of a departure is 0.01. The values exceeding these critical levels were deemed departures.

To obtain the power functions of the $\bar{\chi}^2$ statistic, we used the approach of Bartholomew (1961), as summarized by Barlow et al. (1972,

Section 3.4). The power function for alternatives within H_1 depends on the spacing of the means. Let

$$(5) \qquad \Delta^2 = \sum_{i=1}^{k} (\mu_i - \mu)^2,$$

where $\mu = \sum_{i=1}^{k} \mu_i/k$, $\mu_i = i\Delta c + \theta$, $i = 1,\cdots,k$, θ is any value and c is a normalizing constant ensuring that (5) holds. (Note that small Δ values correspond to closely spaced means.) For three treatment groups the power is a maximum when the means are equally spaced and a minimum when two of the means are equal. Thus, we denote by 1,2,3 and 1,2,2 the alternative spacings for which the power function is a maximum and a minimum, respectively. We simulated the results for $\Delta = 1,2,3$ and 4.

For the null distribution, simulations were performed for $\bar{\chi}^2$ and $\bar{E}^2$ on the nine distributions and for $\bar{\chi}^2_{rank}$ on three treatment groups with $n = 5$, 10, 15, 20, and 25. For the power function, simulations were performed for the 1,2,2 and the 1,2,3 spacing for $\bar{\chi}^2$, $\bar{E}^2$ and $\bar{\chi}^2_{rank}$ on three treatments with $n = 5$ and $\Delta = 1,2,3$ and 4. In the alternative space, the $\bar{\chi}^2_{rank}$ tests were made on all nine distributions, since under H_1, the distribution of the test depends on F. For each simulation the power function was evaluated at five places, corresponding to $\alpha = .100$, .050, .025, .010 and .005 under H_0. All simulations consist of 10,000 repetitions, with use of random number generation routines from the IMSL library.

3.2 *The Robustness of the* $\bar{\chi}^2$, $\bar{E}^2$ *and* $\bar{\chi}^2_{rank}$ *Tests.* For this paper a distribution will be said to be robust at a given sample size if all departures from robustness occur at smaller sample sizes. Minimum sample sizes for robustness were determined for each distribution. (In three instances a departure was noted for a given sample size with two successive nondepartures at lower sample sizes. In these cases, the lower sample size with no departure was used.) The distributions were then dichotomized as either flat (kurtosis ≤ 4) or peaked (kurtosis ≥ 6), with the t_3 distribution receiving special note (Table 1). For flat distributions the $\bar{\chi}^2$ statistic is generally robust at sample sizes as small as 5. Further, three of the four departures are conservative; that is, the simulated alpha level is less than the nominal alpha level. On the other hand, for peaked distributions larger sample sizes are required for the $\bar{\chi}^2$ statistic

Table 1. Median minimum sample size for robustness of $\bar{\chi}^2$, $\bar{E}^2$ and $\bar{\chi}^2_{rank}$ for flat and peaked distributions*

	$\bar{\chi}^2_{rank}$	$\bar{\chi}^2$			$\bar{E}^2$		
α		Flat[a]	Peaked[b]	t_3	Flat[a]	Peaked[b]	t_3
.1000	5	5	5	25	5	5	5
.0500	5	5	5	15	5	5	5
.0250	5	5	12.5	5	5	7.5	15
.0100	10	5	15	>25	5	5	25
.0050	10	5	12.5	>25	5	5	5
Departures/ Distribution	2	1	5	15	0.5	1	4
Global[c]	15	5	10	>25	-	-	-

*For this table only, t_3 has not been included with the remaining peaked distributions.

[a]Uniform, normal, NM (.1,4), NM (.5,2).

[b]Double exponential, lognormal, exponential, NM (.9,4).

[c]The KS, L_1, L_2, and χ^2 test statistics yield nearly identical results so only the overall median for the global tests is given here.

to be robust, particularly in the extreme tail $(\alpha < .01)$ of the distribution. Of the 35 departures all but five are anticonservative, the five exceptions being at the .10 and .05 α-levels for t_3. The t_3 distribution demonstrates nonrobust behavior in the extreme tail, even for samples of size 25.

The departures seen for KS, L_1, L_2 and χ^2 corroborate the α-level specific departure data. Flat distributions are robust for n as small as 5 whereas peaked distributions are robust for $n \leq 20$, with the exception of t_3, which is not robust even for $n = 25$.

The $\bar{E}^2$ test proved to be much more robust than the $\bar{\chi}^2$ test (Table 1), with only 10 departures, compared to 39 for the $\bar{\chi}^2$ test. Here all 10 departures were conservative, suggesting that this test could be used even when nonrobust behavior occurs. No summary statistics assessing the robustness of $\bar{E}^2$ were undertaken since its distribution is a function of the sample size, complicating the choice of partitions of the sample space.

Critical values for the $\bar{\chi}^2_{rank}$ test were determined using the $\bar{\chi}^2$ approximation. The test is then conservative in the extreme tail. When $n = 5$, the simulated alpha levels corresponding to the nominal levels .01 and .005 are .0071 and .0027, respectively. For $n = 10$, however, the simulated alpha levels improve to .0094 and .0040, respectively. For $n = 5$, exact critical levels were determined. This leads to a robust test as one would expect.

3.3 *The Power of* $\bar{\chi}^2$, $\bar{E}^2$ *and* $\bar{\chi}^2_{rank}$ *in the Alternative Space.* Before comparing the tests with respect to power, we will describe the results of each test separately. The theoretical powers for each Δ,α pair for the $\bar{\chi}^2$ test statistic were computed with the power function equation given in Bartholomew (1961). For the 1,2,3 spacing, Table

2 lists the powers for $\Delta = 1$ and 2; for $\Delta = 3$ the powers range from .62 to .95, and for $\Delta = 4$ all powers exceed .90. The powers for the 1,2,2 spacing are all within .036 of the 1,2,3 spacing powers at the levels chosen. The simulation results are similar for the 1,2,2 and the 1,2,3 spacings. Hence, unless explicitly

Table 2. Simulated minimum and maximum relative powers for flat and peaked distribution with 1,2,3 spacing.

α	Exact power	Minimum	Maximum	Number of departures
		Flat[a]		
		$\Delta = 1$		
.1000	.370	0.99	1.02	0
.0500	.244	0.95	1.01	1
.0250	.156	0.96	1.03	0
.0100	.084	0.92	1.08	0
.0050	.052	0.92	1.06	0
		$\Delta = 2$		
.1000	.735	0.99	1.01	0
.0500	.605	0.98	1.02	0
.0250	.481	0.98	1.02	0
.0100	.340	0.97	1.03	0
.0050	.254	0.96	1.03	0
		Peaked[b]		
α	Exact power	Minimum	Maximum	Number of departures
		$\Delta = 1$		
.1000	.370	0.90	0.99	4
.0500	.244	0.84	0.96	4
.0250	.156	0.81	0.96	3
.0100	.084	0.86	1.02	1
.0050	.052	0.93	1.10	0
		$\Delta = 2$		
.1000	.735	1.00	1.02	2
.0500	.605	0.99	1.01	0
.0250	.481	0.96	1.00	3
.0100	.340	0.89	0.99	3
.0050	.254	0.85	0.99	3

[a]Uniform, normal, NM (.1,4), NM (.5,2)

[b]Double exponential, lognormal, exponential, NM (.9,4), t_3

mentioned, all comments below refer to the 1,2,3 spacing data.

The $\bar{\chi}^2$ test power function is robust for flat distributions at all levels of Δ, with only two departures occurring in the 80 Δ,α-level by distribution combinations. For peaked distributions, however, 34 departures occur in the 100 combinations. For closely spaced means ($\Delta = 1$ or 2), peaked distributions have relative power less than 1, with 21 and 23 departures being low (Table 2). For widely spaced means ($\Delta = 3$ or 4), however, a reversal occurs and 9 of 11 departures have relative power greater than 1. The behavior is most pronounced in the t_3 distribution. For $\Delta \le 2$, the distribution has the minimum relative power in 9 of 10 instances, while for $\Delta \ge 3$ it has the maximum relative power in 9 of 10 instances, accounting for 15 of the 34 departures seen. The minimum relative powers of the 1,2,2 spacing simulation are slightly lower than they are for the 1,2,3 spacing, but the general observations presented above hold here as well.

The simulations on $\bar{E}^2$ disclose a strong connection between the power function and the kurtosis of the underlying distribution. The peaked distributions, except for the lognormal, lead to conservative tests. For $\alpha \le .025$ the relative power is at most 0.81. For $\Delta = 1$ and 2, the power functions increase as the kurtosis increases (see Figure 1). It was expected that the lognormal would have higher power than the double exponential for all values of alpha, as they both have kurtoses of 6 and the lognormal is skewed. However, the lognormal has lower power in the extreme tail. For $\Delta = 3$ and 4, the leptokurtic distributions (distributions with kurtosis greater than 3) have higher power for small alpha but lower power for large alpha levels. For $\Delta = 3$, the switch occurs at $\alpha = .05$, for $\Delta = 4$ it occurs at $\alpha = .025$.

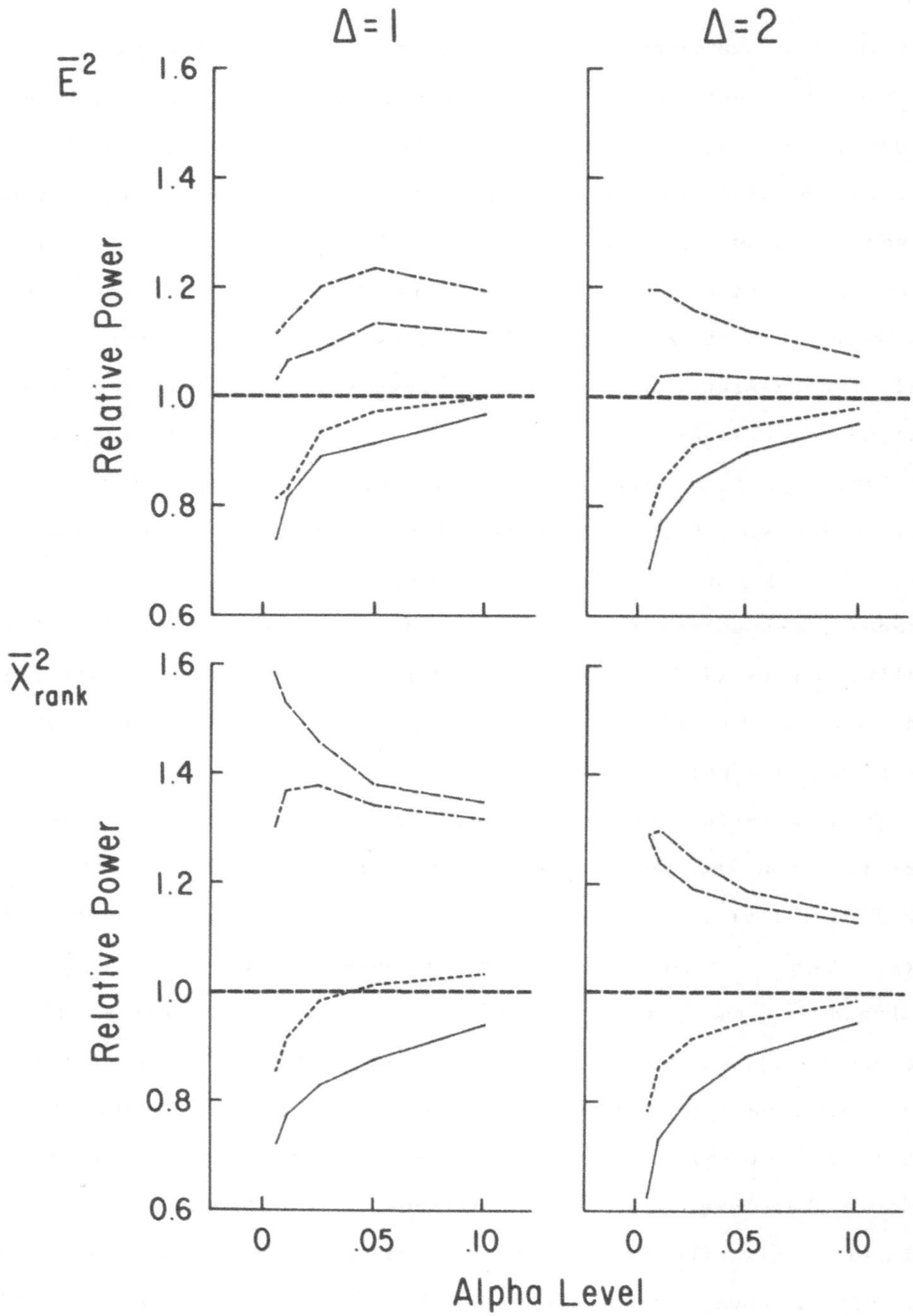

Figure 1. Relative powers for $\bar{E}^2$ and $\bar{\chi}^2_{rank}$. Normal distribution-solid line, NM(.5,2) - short dash line, exponential - long dash line, t_3 - mixed dash line.

The simulations on $\bar{\chi}^2_{rank}$ show a pattern somewhat similar to that seen for $\bar{E}^2$. For $\Delta = 1$, the power function is larger for distributions with larger kurtoses, except for the exponential and double exponential, which have larger power than their kurtoses would indicate. For $\Delta \geq 2$, the power functions align by kurtoses (Figure 1).

A comparison of the three tests shows that $\bar{\chi}^2_{rank}$ and $\bar{E}^2$ improve with increasing kurtosis, relative to $\bar{\chi}^2$. Denote by CER a test comparison in which, for a given situation, the $\bar{\chi}^2$ test (C) has the highest power and the $\bar{\chi}^2_{rank}$ test (R) the lowest power for at least three of the five alpha levels used. (The $\bar{E}^2$ test is denoted by E.) Define the other trigrams in a similar fashion. As the kurtosis increases or Δ decreases, the test orderings change from CER to CRE to RCE, and then to REC, that is, $\bar{\chi}^2_{rank}$ outperforms $\bar{E}^2$ and then $\bar{\chi}^2$, and finally $\bar{E}^2$ outperforms $\bar{\chi}^2$ (see Table 3). This rule is followed perfectly by the 1,2,3 spacing and with minor violations by the 1,2,2 spacing, which also shows less tendency to depart

Table 3. Power function orderings for $\bar{\chi}^2$, $\bar{E}^2$ and $\bar{\chi}^2_{rank}$.

Distribution	Skewness	Kurtosis	(a) 1,2,3 spacing, Δ = 1	2	3	4	(b) 1,2,2 spacing, Δ = 1	2	3	4
Uniform	0.0	1.8								
Normal	0.0	3.0			CER				CER	
NM (.1,4)	-0.1	3.3	CRE							
NM (.5,2)	-0.5	3.9		CRE			CRE	CRE		
Double Exp.	0.0	6.0		RCE			RCE	RCE	CRE	
Lognormal	1.3	6.0			CRE		RCE	CRE		
Exponential	2.0	9.0			CRE	CRE				
NM (.9,4)	-1.0	13.0	REC		RCE	CRE	REC		RCE	CRE
t_3	0.0	DNE*			RCE	CRE			CRE	CER

*DNE - does not exist.

from CER. These tests are graphically compared for representative distributions in Figure 2 and are given in Table 4 when F is normal.

Table 4. $\bar{\chi}^2$, $\bar{E}^2$ and $\bar{\chi}^2_{rank}$ relative powers for the 1,2,3 spacing when F is normal.

Δ	Statistic	α .1000	.0500	.0100
	$\bar{\chi}^2$	1.02	1.01	1.08
1	$\bar{E}^2$	.97	.91	.82
	$\bar{\chi}^2_{rank}$	.94	.88	.77
	$\bar{\chi}^2$	1.00	1.00	.97
2	$\bar{E}^2$	.95	.90	.77
	$\bar{\chi}^2_{rank}$	.94	.88	.73
	$\bar{\chi}^2$	1.00	1.00	1.00
3	$\bar{E}^2$	.99	.98	.82
	$\bar{\chi}^2_{rank}$	.97	.92	.77

It is somewhat surprising that the $\bar{E}^2$ test would ever outperform the $\bar{\chi}^2$ test, since knowledge of σ^2 should provide a sharper test for detecting mean orderings in the alternative space. Indeed, for a normal underlying distribution the CER pattern was observed in all cases, with the rank test losing little power to the $\bar{E}^2$ test. To investigate the patterns seen for the normal and t_3 distribution, given 1,2,3 spacing, we generated a sample of size 100 for each of four situations, to which all three tests were applied. The discordances in rejecting the null hypothesis were seen to correlate with

$$(6.6) \qquad \sum_{i=1}^{k} \sum_{j=1}^{n} (X_{ij}-\hat{\mu})^2,$$

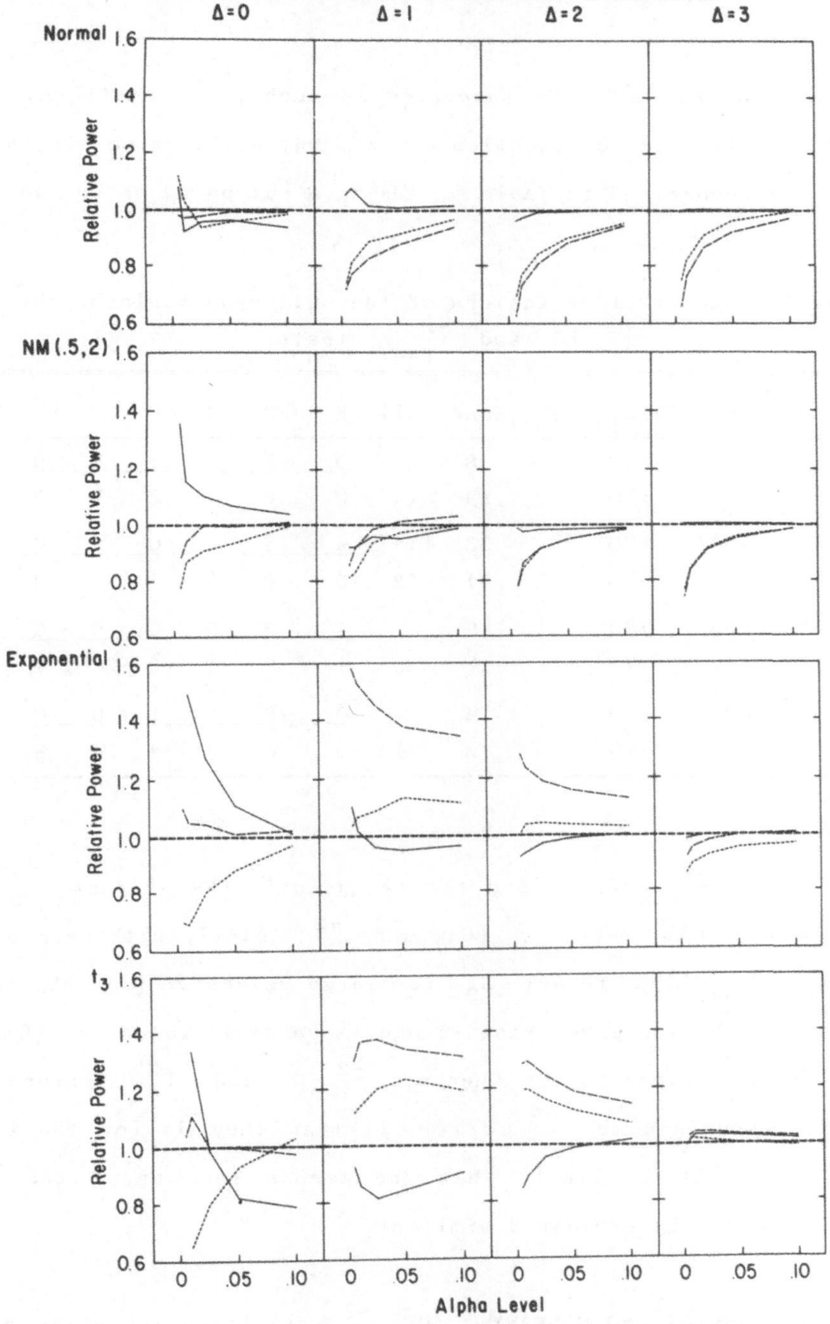

Figure 2. Relative powers for $\bar{\chi}^2$, $\bar{E}^2$ and $\bar{\chi}^2_{rank}$ by distribution. Normal distribution - solid line, NM(.5,2) - short dash line, exponential - long dash line, t_3 - mixed dash line.

the denominator of $\bar{E}^2$. We suspected as much since small values of (6) could easily lead to rejection of $\bar{E}^2$ but not necessarily to $\bar{\chi}^2$. The data are summarized in Table 5. Using a cut point of 70, we see

Table 5. Number of rejections of the null hypothesis by the $\bar{\chi}^2$, $\bar{E}^2$ and $\bar{\chi}^2_{rank}$ tests.

Distri-bution	Δ	α	$\Sigma\Sigma(X_{ij}-\hat{\mu})^2$	None	All	E	ER	R	EC	C	RC	Total
Normal	1	.100	≤ 70	38	9	0	5	2	1	0	0	55
			>70	24	14	0	0	1	2	2	2	45
t_3	1	.100	≤ 70	33	14	4	14	7	0	0	0	72
			>70	11	12	0	0	0	1	1	3	28
Normal	3	.005	≤ 70	6	2	1	3	0	0	0	0	12
			>70	17	36	1	0	1	6	26	1	88
t_3	3	.005	≤ 70	4	7	0	15	1	1	0	0	28
			>70	13	31	0	0	0	4	16	8	72

that $\bar{E}^2$ (E), $\bar{\chi}^2_{rank}$ (R), and the two jointly (ER) reject H_0 for small values of (6), while $\bar{\chi}^2$ (C) and $\bar{\chi}^2$ jointly with each of $\bar{E}^2$ (EC) and $\bar{\chi}^2_{rank}$ (RC) reject H_0 for large values of (6). Moreover, the t_3 distribution gives rise to many more small values of (6) than does the normal distribution; hence, $\bar{\chi}^2_{rank}$ and $\bar{E}^2$ perform much better in comparison to $\bar{\chi}^2$ for t_3 than they do for the normal distribution. It is likely that the trends seen among the other distributions can be explained similarly.

4. DISCUSSION AND SUMMARY. THE $\bar{\chi}^2$ statistic is fairly robust when applied to nonnormal underlying distributions, as noted in Table 2. The flat (kurtosis $\le$ 4) distributions give rise to test statistic distributions that are well approximated by the normal distribution, under both H_0 and H_1. For peaked distributions, convergence is slower and the test is anticonservative, which is undesirable. Under H_1, this class produces lower power than would be expected for closely spaced means and higher power for widely spaced means. The $\bar{E}^2$ statistic is more robust than the $\bar{\chi}^2$ test under H_0, possibly because some $\bar{\chi}^2$ test rejections emerging from heavy-tailed distributions are corrected for the denominator of $\bar{E}^2$. Peaked distributions lead to conservative tests, however. Two facts can be noted about the $\bar{E}^2$ statistic for closely spaced alternatives. First, the power functions increase in concert with the kurtosis of the underlying distribution. Second, the $\bar{E}^2$ test has larger power than the $\bar{\chi}^2$ test for peaked distributions. The $\bar{\chi}^2_{rank}$ test, however, often significantly outperforms that of the $\bar{E}^2$ test and never seems to perform dramatically worse. Thus, given the observation above, the anticonservatism of the $\bar{\chi}^2$ test and the fact that the variance is seldom known in practice, the $\bar{\chi}^2_{rank}$ test becomes an attractive alternative test in studies of order restricted inference.

This paper has dealt primarily with the performance of $\bar{\chi}^2$, $\bar{E}^2$ and $\bar{\chi}^2_{rank}$ for a simple order alternative on three treatments with equal samples of size 5. The performance characteristics indicate that the rank test is preferable to the $\bar{\chi}^2$ or $\bar{E}^2$ test when the underlying distribution is heavy-tailed.

ACKNOWLEDGEMENTS. We are grateful to Dr. Farroll T. Wright and Dr. Diane Fairclough for their helpful suggestions, and to Nancy Hendren for preparation of the manuscript.

REFERENCES

Barlow, R.E., Bartholomew, D.J., Bremner, J.M. & Brunk, H.D. (1972). *Statistical Inference Under Order Restrictions*. Wiley: New York.

Bartholomew, D.J. (1961). A test of homogeneity of means under restricted alternatives. *J. Roy. Statist. Soc.* **B23**, 239-281.

Chacko, V.J. (1963). Testing homogeneity against ordered alterntives. *Ann. Math. Statist.* **34**, 945-956.

IMSL, (1985). Int. Math. and Stat. Library, Version 9.2. IMSL, Inc.: Houston, Texas.

Parsons, V.L. (1979). A note on the exact distribution of a nonparametric test statistic for ordered alternatives. *Ann. Statist.* **7**, 454-458.

Parsons, V.L. (1981). Small sample distribution for a nonparametric test for trend. *Commun. Statist.* **B** (3), 289-302.

Michael J. Schell
Biostatistics Section
St. Jude Children's Research Hospital
332 N. Lauderdale, P.O. Box 332
Memphis, TN 38101

Bahadur Singh
Department of Mathematical Sciences
Memphis State University
Memphis, TN 33152

MULTIPLE CONTRAST TESTS FOR TESTING AGAINST A SIMPLE TREE ORDERING[1]

Hari Mukerjee
University of California, Davis

Tim Robertson
The University of Iowa

F.T. Wright
University of Missouri-Rolla

AMS 1980 subject classifications: *Primary* 62F03, *Secondary* 62H15.
Key words and phrases: order restricted tests, multiple contrast tests, simple tree ordering, power of tests.

ABSTRACT

Consider k treatments with means $\mu_1, \mu_2, \cdots, \mu_k$, and a control with mean μ_0. Consider the hypotheses $H_0: \mu_0 = \mu_1 = \cdots = \mu_k$, $H_1: \mu_0 \le \mu_i$ for all $1 \le i \le k$, and $H_2: \mu_o > \mu_i$ for some $1 \le i \le k$. There are several procedures for testing H_0 versus H_1-H_0 (H_1 but not H_0). Testing H_1 versus H_2 would provide a check on the ordering assumption underlying these one-sided procedures. For the latter case, this paper compares the likelihood ratio test, single contrast tests based on several criteria of optimality, and a new multiple contrast test based on the maximum of the contrasts of the vector of observed means against the corners of the Fenchel dual of the

(1) This research was sponsored by the Office of Naval Research under ONR contracts N00014-80-C0321 and N00014-80-C0322.

cone H_1, which is in a sense dual to Dunnett's test for testing H_0 versus H_1-H_0 in the same setting. Based on ease of implementation, comparison of powers and computability of powers, the multiple contrast procedure appears to be the most attractive.

1. INTRODUCTION. We consider the experimental situation in which one wishes to compare several treatments with a control or standard. In many situations it is believed that the treatments are at least as effective as the control, and procedures have been developed which make use of this information in the inference. For instance, if the control and treatment populations are normal with a common unknown variance, then Dunnett's (1955) one-sided procedures can be used to test homogeneity of the means with the alternatives restricted by the assumption that the treatment means are at least as large as the control mean. Dunnett (1955) also provides a multiple comparison procedure for this one-sided problem. Brunk's (1955) work gives the restricted maximum likelihood estimates in this setting and Bartholomew's (1959, 1961) $\bar{\bar{E}}^2$ test is the likelihood ratio test of homogeneity with such restricted alternatives. Bartholomew calls the ordering, which requires each treatment mean to be at least as large as the control, a simple tree ordering. Using the work of Abelson and Tukey (1963), Barlow et al. (1972, p. 186) discuss optimal contrast tests for these hypotheses. Mukerjee, Robertson and Wright (1985) studied a test based on the maximum of orthogonal contrasts which is a competitor to the tests proposed by Dunnett and Bartholomew as well as the optimal contrast test.

In some cases it may be desirable to have a test to check the assumption that the treatments are at least as effective as the control. To be specific, in the normal means setting, let μ_0 denote the control mean; let μ_i, $i = 1,2,\ldots,k$, denote the treatment means; and consider $H_0: \mu_0 = \mu_1 = \cdots = \mu_k$, $H_1: \mu_0 \le \mu_i$, for all $i = 1,2,\ldots,k$, and $H_2: \mu_0 > \mu_i$ for some $1 \le i \le k$. Let μ denote the vector of these means in R^{k+1}. The procedures mentioned above provide tests of H_0 versus H_1-H_0, i.e., H_1 holds but H_0 does not. Testing H_1 versus H_2 would provide a check on the ordering assumption underlying these one-sided procedures. A test of H_1 versus H_2 would also be useful in any situation in which one wishes to determine if a collection of means are all larger than a given mean and the probability of rejecting H_1, when it is true, needs to be controlled. Robertson and Wegman (1978) developed the likelihood ratio test of H_1 versus H_2, and in fact, they considered arbitrary partial orders. We denote their test statistic by $\bar{E}^2_{12}$. They have shown that H_0 is least favorable within H_1 and that, under H_0, the distribution of $\bar{E}^2_{12}$ is a mixture of beta distributions. If the sample sizes are unequal, the mixing coefficients are quite complex even for moderate k. While the complexity of the null distribution of $\bar{E}^2_{12}$ is a drawback, even more important is the fact that the power function is extremely complicated outside of H_1. (See Robertson and Wright (1985) and the discussion in Section 3.) Of course, this is serious if one wishes to design an experiment to achieve a specified power at a particular alternative.

The same difficulty is encountered in testing H_0 versus H_1-H_0, i.e., the likelihood ratio test (based on the statistic $\bar{E}^2_{01}$) also has a complex power function outside of H_0. However, the orthogonal contrast test proposed by Mukerjee, Robertson and Wright (1985) has a much more tractable distribution under the null and alternative hypotheses, while its power function is comparable to that of $\bar{E}^2_{01}$. In Section 3 it is observed that no suitable orthogonal contrast test exists for testing H_1 versus H_2, and so a natural multiple contrast test is proposed for testing H_1 versus H_2. The procedure, which is in a sense dual to Dunnett's test of H_0 versus H_1-H_0, rejects H_1 for large values of

$$(1.1) \qquad T_D = \max_{1\le i\le k} (\bar{X}_0-\bar{X}_i)/[S(n_0^{-1}+n_i^{-1})^{1/2}],$$

where $\bar{X} = (\bar{X}_0,\bar{X}_1,\ldots,\bar{X}_k)$ is the vector of sample means, $n_0,n_1,\ldots,n_k$ are the sample sizes, and S^2 is the usual pooled estimate of σ^2. Conditioning on S, we see that H_0 is also least favorable within H_1 for this multiple contrast test. Under H_0, T_D has the same distribution as Dunnett's test statistic with $\bar{X}_0$ and $\bar{X}_i$ interchanged. The critical values of T_D are tabled, and, if the degrees of freedom are large enough that the case of a known variance provides a reasonable approximation, then the powers can be obtained by a single numerical integration.

In Section 2 an optimal contrast test is obtained for the case of a known variance, and the studentized statistic, T_S, is considered when the variance is unknown. Of course, the distributions of T_S under the null and alternative hypotheses are quite straightforward. In Section 3 the powers of T_S, T_D and the likelihood ratio test are compared. The single contrast test has good power in the "middle" of the alternative region, but its power is unacceptably low at other points. Hence, if the alternatives cannot be narrowed down further than $\mu \in H_1$, a test based on T_S cannot be recommended. Dunnett's procedure was compared with the likelihood ratio test for H_0 versus

H_1-H_0 in Robertson and Wright (1985) and Mukerjee, Robertson and Wright (1985). Their powers are very much alike for $k = 2$ and 3. For $k = 4$, the likelihood ratio test has a slightly larger power, and this difference in powers increases with k. For testing H_0 versus H_1-H_0, Dunnett's procedure has greater power than the likelihood ratio test in the center of H_1, but falls off rapidly, for moderate and large k, as the alternative moves away from the center. In the case considered here, the likelihood ratio test has larger power in the middle of the alternatives, but Dunnett's procedure is better at points away from the middle. Because of its ease of implementation, Dunnett's procedure seems to be a viable alternative to the likelihood ratio test in the special case of testing H_1 versus H_2. (See Section 3 for specific power comparisons.) This is particularly interesting since Robertson and Wegman (1978) compared the likelihood ratio test with the analogous multiple contrast test for testing $\mu_1 \le \mu_2 \le \cdots \le \mu_k$ versus $\mu_i > \mu_{i+1}$ for some i and found that the likelihood ratio test is clearly superior. Undoubtedly, the difference is due to the fact that the simple tree cone, H_1, is much larger than the cone for a total order, i.e., $\{x \in R^k: x_1 \le x_2 \le \cdots \le x_k\}$.

2. A SINGLE CONTRAST TEST. Suppose one has independent random samples from $k+1$ normal populations with a common variance σ^2 and that one wishes to test H_1 versus H_2. As in the Introduction, we denote the sample means by $\overline{X}_0, \overline{X}_1, \ldots, \overline{X}_k$, the sample sizes by $n_0, n_1, \ldots, n_k$, the degrees of freedom by $m = n_0 + n_1 + \ldots + n_k-k-1$, and the pooled estimate of σ^2 by S^2. Of course, S^2 is independent of the vector of sample means and $mS^2/\sigma^2 \sim \chi^2(m)$. We derive optimal contrast tests for the case in which σ^2 is known and studentize the test statistics for the unknown variance case. The case $k = 1$ has been studied thoroughly, so we suppose $k \geq 2$. For $i = 1,2,\dots,k$, let $w_i = n_i/\sigma^2$. (If the variances are known, then that which follows also holds if the ith population has variance σ_i^2 and $w_i = n_i/\sigma_i^2$.) For $c \in R^{k+1}-\{0\}$, we consider the contrast test which rejects H_1 for large values of

$$T_c = \sum_{i=0}^{k} w_i c_i \bar{X}_i .$$

For a given critical value, t, the power function is given by

$$\pi_c(\mu) = 1-\Phi\{[t-(c,\mu)_w]/\|c\|_w\},$$

where $(c,\mu)_w = \sum_{i=0}^{k} w_i c_i \mu_i$, $\|c\|_w^2 = (c,c)_w$, and Φ is the cdf of the standard normal distribution. Consider a fixed c. If there is a $\mu \in H_1$ (We also let H_1 denote the subset of R^{k+1} which satisfies the hypothesis H_1.) then, because $b\mu = (b\mu_0, b\mu_1, \dots, b\mu_k) \in H_1$ for $b \geq 0$, the significance level is $\sup_{\mu \in H_1} \pi_c(\mu) = 1$ for all $t \in R$. Hence, we restrict attention to c with $(c,\mu)_w \leq 0$ for all $\mu \in H_1$. Now H_1 is a closed, convex cone, i.e., it is a closed subset of R^{k+1}; if $x,y \in H_1$ then $\lambda x + (1-\lambda)y \in H_1$ for $o \leq \lambda \leq 1$; and if $x \in H_1$ and $b \geq 0$, then $bx \in H_1$. The dual of H_1 (sometimes called the polar cone) is defined by

$$H_1^{*w} = \{y \in R^{k+1} : (x,y)_w \leq 0 \text{ for all } x \in H_1\}.$$

Thus we assume

(2.1) $$c \in H_1^{*w}.$$

Dykstra (1984) gives an excellent discussion of statistical applications of dual cones, from which it easily follows that

$$H_1^{*w} = \{y \in R^{k+1}: \sum_{i=0}^{k} w_i y_i = 0, \ y_i \le 0 \ \text{ for } \ i = 1,2,\dots,k\}.$$

Hence, $\sum_{i=0}^{k} w_i c_i = 0$. Thus, if $\Phi(z_\alpha) = 1-\alpha$, then $t = z_\alpha \|c\|_w$ gives a test of size α, and the power function becomes

$$\pi_c(\mu) = 1-\Phi[z_\alpha-(c,\mu)_w/\|c\|_w].$$

Next, we consider different optimality criteria and derive expressions for the corresponding optimal contrast vectors. First, we review some basic ideas concerning projections. (The reader may wish tc consult Chapter 7 of Barlow et al. (1972).) The projection with respect to $\|\cdot\|_w$ onto a closed, convex set $C \subset R^{k+1}$ exists and is unique. We denote it by $E_w(\cdot|C)$. If C is also a cone, then $E_w(x|C)$ is characterized by

(2.2) $$E_w(x|C) \in C, \ (x-E_w(x|C),y)_w \le 0 \ \text{ for all } \ y \in C, \ \text{ and } \quad (x-E_w(x|C),E_w(x|C))_w = 0;$$

$E_w(bx|C) = bE_w(x|C)$ for $b \ge 0$; and, if C contains the constant vectors (i.e., H_0), then

(2.3) $$\sum_{i=0}^{k} w_i E_w(x|C)_i = \sum_{i=0}^{k} w_i x_i.$$

Applying (2.2) and the definition of H_1^{*w}, we see that

$$E_w(x|H_1^{*w}) = x - E_w(x|H_1) \quad \text{and} \tag{2.4}$$
$$\|E_w(x|H_1^{*w})\|_w^2 = \|x\|_w^2 - \|E_w(x|H_1)\|_w^2.$$

Finally, $E_w(x|H_0)$ is a constant vector with each component equal to $\tilde{x} = \sum_{i=0}^{k} w_i x_i / \sum_{i=0}^{k} w_i$.

With $\mu \notin H_1$ fixed, we obtain the contrast test which maximizes the power at μ, i.e., c maximizes $(d,\mu)_w/\|d\|_w$ over all $d \in H_1^{*w} - \{0\}$. Since $\tilde{d} = 0$, equivalently, c maximizes

$$\rho(d,\mu) \equiv \sum_{i=0}^{k} w_i(d_i-\tilde{d})(\mu_0-\tilde{\mu})/(\|d-\tilde{d}\|_w\|\mu-\tilde{\mu}\|_w)$$

over all $d \in H_1^{*w}-\{0\}$. Because $\mu \notin H_1$, there exists $j \in \{1,2,\ldots,k\}$ with $\mu_j < \mu_0$. Now, if $c_0 = w_0^{-1}$, $c_j = -w_j^{-1}$, and $c_i = 0$ for $i \notin \{0,j\}$, then $\rho(c,\mu) > 0$. Using the convention that $\rho(c,0) = \rho(0,\mu) = 0$, the maximization problem is unchanged if $H_1^{*w}-\{0\}$ is replaced by H_1^{*w}. For $A,B \subset R^{k+1}$, let $A \oplus B = \{x+y : x \in A,\ y \in B\}$. Clearly, $H_1^{*w} \oplus H_0$ is a closed, convex cone which contains the constant vectors. Furthermore, c maximizes $\rho(d,\mu)$, $d \in H_1^{*w} \oplus H_0$, if and only if $c-\tilde{c}$ maximizes $\rho(d,\mu)$, $d \in H_1^{*w}$. Thus we seek c which maximizes $\rho(d,\mu)$ subject to $d \in H_1^{*w} \oplus H_0$ and we will base the contrast test on $c-\tilde{c}$. Applying (ii) of Corollary E of Barlow et al. (1972, p. 320), we see that $c = E_w(\mu|H_1^{*w} \oplus H_0)$ solves this maximization problem. Using (2.2) it is easily shown that $E_w(\mu|H_1^{*w} \oplus H_0) = E_w(\mu|H_1^{*w})+\tilde{\mu}$, and applying the characterization of H_1^{*w}, we see that $\sum_{i=0}^{k} w_i(E_w(\mu|H_1^{*w})+\tilde{\mu})/\sum_{i=0}^{k} w_i = \tilde{\mu}$. Hence, $c-\tilde{c} = E_w(\mu|H_1^{*w})$. The power function of the test with size α and $c = E_w(\mu|H_1^{*w})$ is

$$1-\Phi[z_\alpha-(E_w(\mu|H_1^{*w}),\mu)_w/\|E_w(\mu|H_1^{*w})\|_w],$$

which becomes $1 = \Phi[z_\alpha-\|E_w(\mu|H_1^{*w})\|_w]$ after applying (2.2). We have proved the following

THEOREM 2.1. *Let* $\mu \notin H_1$. *The contrast test with maximum power at* μ *is determined by* $c = E_w(\mu|H_1^{*w}) = \mu-E_w(\mu|H_1)$, *and its power function is given by*

$$\pi_c(\mu) = 1-\Phi[z_\alpha- \|E_w(\mu|H_1^{*w})\|_w].$$

Since the optimal c depends on the unknown μ, one could estimate it using $E_w(\overline{X}|H_1^{*w}) = \overline{X}-E_w(\overline{X}|H_1)$. However, this adaptive choice of c gives rise to precisely the likelihood ratio test statistic $\sum_{i=0}^{k} w_i[\overline{X}_i-E_w(\overline{X}|H_1)_i]^2$ developed by Robertson and Wegman (1978) for the case of known variances. We state this result as a theorem:

THEOREM 2.2. *If* $c = E_w(\overline{X}|H_1^{*w})$ *then*

$$T_c = \sum_{i=0}^{k} w_i[\overline{X}_i-E_w(\overline{X}|H_1)_i]^2.$$

PROOF. On applying (2.2),

$$T_c = \sum_{i=0}^{k} w_i[\bar{X}_i - E_w(\bar{X}|H_1)_i]\bar{X}_i = \sum_{i=0}^{k} w_i[\bar{X}_i - E_w(\bar{X}|H_1)_i]^2.$$

Thus the likelihood ratio test is an adaptive contrast test. This was observed earlier for testing H_0 versus H_1-H_0 with an arbitrary partial order (Theorem 4.3 of Barlow et al. (1972)). It should be noted that the discussion given thus far in this section also holds for arbitrary partial orders.

Next we consider more global optimality criteria. Abelson and Tukey (1963), in related testing situations, obtained the contrast tests which maximize the minimum of the power over points at a distance δ from the null hypothesis. Using their criterion, we wish to obtain $c \in H_1^{*w}-\{0\}$ so that

$$\inf\{\pi_c(\mu): \|\mu-E_w(\mu|H_1)\|_w = \delta\} \geq \inf\{\pi_{c'}(\mu): \|\mu-E_w(\mu|H_1)\|_w = \delta\}$$

for all $c' \in H_1^{*w}-\{0\}$. However, it follows from the two lemmas given below that, for $c \in H_1^{*w}-\{0\}$ and $\delta > 0$, $\inf\{\pi_c(\mu): \|\mu-E_w(\mu|H_1)\|_w = \delta\} = 0$. Thus other criteria must be considered. The proofs of the two lemmas are given in the Appendix.

LEMMA 2.3. *If* $c \in H_1^{*w}-\{0\}$, $\delta > 0$, $k \geq 2$, *then there exists* $\mu \in H_1$ *with* $\|\mu-E_w(\mu|H_1)\|_w = \delta$ *and* $(c, E_w(\mu|H_1))_w < 0$.

LEMMA 2.4. *For* $b \geq -1$ *and* $\mu \in R^{k+1}$, $\|\mu_b-E_w(\mu_b|H_1)\|_w = \|\mu-E_w(\mu|H_1)\|_w$ *with* $\mu_b = \mu+bE_w(\mu|H_1)$.

Following Schaafsma and Smid (1966), we consider the contrast test that minimizes the maximum "shortcoming" among all contrast tests. Recall that for a given $\mu \notin H_1$, the contrast test with maximum power at μ is given by setting $c = \mu - E_w(\mu|H_1)$ and has power $1-\Phi[z_\alpha - \|\mu - E_w(\mu|H_1)\|_w]$. Thus, for any contrast test based on a fixed contrast c, its shortcoming at μ is

$$\Phi[z_\alpha - (c,\mu)_w/\|c\|_w] - \Phi[z_\alpha - \|\mu - E_w(\mu|H_1)\|_w].$$

If there is no further constraint on μ other than $\mu \notin H_1$, we see from the preceding analysis that the supremum of the shortcoming can be made as large as $1-\Phi(z_\alpha - \delta)$ for each $\delta > 0$. Thus the supremum is 1, and this criterion is not appropriate here. (If $\mu \notin H_1$ and μ is further constrained to be at a distance δ from H_0, then the supremum is $1-\Phi(z_\alpha-\delta)$, which does not depend on c. Hence, this does not provide a choice for c.)

In both instances above, the vector $\mu + bE(\mu|H_1)$ remains at a fixed distance from H_1, but it is moving away from H_0 as b increases. This observation suggests another optimality criterion: Find the contrast test which maximizes the minimum power over all $\mu \notin H_1$ with

$$\Delta^2(\mu) = \sum_{i=0}^{k} w_i(\mu_i - \tilde{\mu})^2 = \delta^2 \quad \text{for some } \delta > 0.$$

Let $b_0 = 0$ and $b_i = (w_0^{-1} + w_i^{-1})^{1/2}$ for $i = 1,2,\ldots,k$ and set

$$c_i^* = \left[\sum_{j=1}^{k} w_j b_j \Big/ \sum_{j=0}^{k} w_j\right] - b_i \quad \text{for } i = 0,1,\ldots,k. \tag{2.5}$$

Denote the contrast statistic based on the contrast vector c^* by

T_S'. If the treatment sample sizes are equal, then $b_1 = b_2 = \cdots = b_k$ and $c_i^* < 0$ for $i = 1,2,\ldots,k$. Hence, for this important case $c^* \in H_1^{*w}$. While c^* is not always in H_1^{*w} for unequal sample sizes, we will show that when it is, T_S' is an optimal contrast statistic.

THEOREM 2.5. *Let* $\delta > 0$ *and* $c^* \in H_1^{*w}$. *The contrast test which rejects for large values of* T_S' *maximizes the minimum of the power over all* $\mu \notin H_1$ *with* $\Delta(\mu) = \delta$. *Furthermore, these contrast coefficients are unique up to a positive multiplier.*

PROOF. The desired contrast vector, c, maximizes $\inf\{1-\Phi[z_\alpha -(d,\mu)_w/\|d\|_w] : \mu \notin H_1, \Delta(\mu) = \delta\}$ subject to $d \in H_1^{*w}-\{0\}$, or, since $\tilde{d} = 0$, c maximizes $\inf_{\mu \notin H_1} \rho(d,\mu)$ subject to $d \in H_1^{*w}-\{0\}$. If we let $b = -c$, then b solves

$$\text{(2.6)} \qquad \text{minimize } \sup_{\mu \notin H_1} \rho(d,\mu) \text{ subject to } d \in -H_1^{*w}-\{0\},$$

where $-A = \{-x : x \in A\}$. Note that $-(H_1^{*w} \oplus H_0) \subset H_1$ from the characterization of H_1^{*w} and the definition of H_1. We will see that for every $d \in -H_1^{*w}-\{0\}$ the supremum in (2.6) is $\rho(d,\mu_d)$ for some $\mu_d \in \partial H_1 - H_0$, where ∂H_1 is the boundary of H_1. Because of the continuity of $\rho(d,\cdot)$, the supremum in (2.6) could be taken over $\mu \notin H_1^0 \cup H_0$, where A^0 denotes the interior of A. If $\mu \in \partial H_1$, then $(-d,\mu)_w \le 0$ and so $\rho(d,\mu) \ge 0$. For $\mu \in H_1^{*w} \oplus H_0$ we have $\rho(d,\mu) \le 0$. Thus, the supremum in (2.6) could be restricted to $\mu \notin H_1^0 \cup (H_1^{*w} \oplus H_0)$, and b solves

$$\text{(2.7)} \qquad \text{minimize } \sup_{\mu \notin H_1^0 \cup (H_1^{*w} \oplus H_0)} \rho(d,\mu) \text{ subject to } d \in -H_1^{*w}-\{0\}.$$

Furthermore, if $\mu \notin H_1^{*w} \oplus H_0$, then $E_w(\mu|H_1) \in H_0$ (This follows from an easy contradiction argument.) and so $\Delta(E_w(\mu|H_1)) > 0$. Applying (2.3) and (2.4), we see that $0 < \Delta(E_w(\mu|H_1)) = \|E_w(\mu|H_1)-\tilde{\mu}\|_w = \|E_w(\mu-\tilde{\mu}|H_1)\|_w \le \|\mu-\tilde{\mu}\|_w = \Delta(\mu)$, the last inequality following from the norm reducing property of projections. For fixed $d \in -H_1^{*w}-\{0\} \subset H_1$, $\rho(d,\mu) = (\|d\|_w\|\mu-\tilde{\mu}\|_w)^{-1}(d,\mu)_w \le (\|d\|_w\|\mu-\tilde{\mu}\|_w)^{-1}(d,E_w(\mu|H_1))_w$ by (2.2), and the last expression is nonnegative because $d \in -H_1^{*w}$. Hence, $\rho(d,\mu) \le \rho(d,E_w(\mu|H_1))$ for $\mu \notin H_1^{*w} \oplus H_0$. Therefore b solves

$$\text{minimize } \sup_{\mu \notin H_1^0 \cup (H_1^{*w} \oplus H_0)} \rho(d,E_w(\mu|H_1)) \quad \text{subject to} \quad d \in -H_1^{*w}-\{0\},$$

or equivalently,

$$\text{minimize } \sup_{\mu \in \partial H_1 - H_0} \rho(d,\mu) \quad \text{subject to} \quad d \in -H_1^{*w}-\{0\}.$$

The boundary of H_1 is the union of $A_i = \{(x_0,x_1,\ldots,x_k) \in H_1 : x_0 = x_i\}$ for $i = 1,2,\ldots,k$. Because of the convention $\rho(d,0) = 0$, we seek b which solves

$$\text{(2.8)} \qquad \text{minimize } \max_{1 \le i \le k} \max_{\mu \in A_i} \rho(d,\mu) \quad \text{subject to} \quad d \in -H_1^{*w}-\{0\}.$$

Each A_i is a closed, convex cone in R^{k+1} containing the constant functions and $\rho(d,\mu) \ge 0$ for every $\mu \in A_i$. By Corollary E, p. 320 of Barlow et al. (1972), $\max_{\mu \in A_i} \rho(d,\mu) = \rho(d,E_w(d|A_i))$, and by (2.2) this can be written as $\|E_w(d|A_i)\|_w/\|d\|_w$. Restricting attention to d with $\|d\|_w = 1$, b solves

$$\text{minimize } \max_{1\le i\le k}\|E_w(d|A_i)\|_w \quad \text{subject to} \quad d \in -H_1^{*w} \quad \text{and} \quad \|d\|_w = 1,$$

or, since $\|d\|_w^2 - \|E_w(d|A_i)\|_w^2 = \|d-E_w(d|A_i)\|_w^2$,

$$\text{maximize } \min_{1\le i\le k}\|d-E_w(d|A_i)\|_w \quad \text{subject to} \quad d \in -H_1^{*w} \quad \text{and} \quad \|d\|_w = 1.$$

Because $\tilde{d} = 0$ and $d_i > 0$ for $i = 1,2,\ldots,k$, $(w_0d_0+w_id_i)/(w_0+w_i)$ $= (-\sum_{j=1}^{k} w_jd_j+w_id_i)/(w_0+w_i) \le 0$ and $d_i = (d_{i0}^*,d_{i1}^*,\ldots,d_{ik}^*)$ with $d_{i0}^* =$ $d_{ii}^* = (w_0d_0+w_id_i)/(w_0+w_1) \le 0$ and $d_{ij}^* = d_j$ for $j \notin \{0,i\}$ is the projection of d onto A_i. Thus, $\|d-E_w(d|A_i)\|_w^2 = w_0[(w_0d_0+w_id_i)/$ $(w_0+w_i)-d_0]^2+w_i[(w_0d_0+w_id_i)/(w_0+w_i)-d_i]^2 = [w_0w_i/(w_0+w_i)](d_0-d_i)^2$, and b solves

(2.9) $$\text{maximize } \min_{1\le i\le k}[w_0w_i/(w_0+w_i)](d_0-d_i)^2 \quad \text{subject to} \quad d \in -H_1^{*w} \quad \text{and} \quad \|d\|_w = 1.$$

Let $d_e = -ec^*$, with c^* defined by (2.5), and note that for $e > 0$, $\tilde{d}_e = 0$, $d_e \in -H_1^{*w}$ by hypothesis, and $\|d\|_w = e\|c^*\|_w > 0$. Thus, we may select e such that $\|d_e\|_w = 1$.

We now show that d_e chosen above is the unique solution to (2.9), which implies that c^* is the unique, up to a positive multiplier, contrast vector which is being sought. Note that, if $d_e = (d_{e0},d_{e1},\ldots,d_{ek})$, then $w_0w_i(d_{e0}-d_{ei})^2/(w_0+w_i) = e^2$ for $i = 1,2,\ldots,k$. Hence, if $z \in -H_1^{*w}$, $\|z\|_w = 1$, and $\min_{1\le i\le k} w_0w_i(z_0-z_i)^2/(w_0+w_i) \ge e^2$, then $z_i-z_0 \ge d_{ei}-d_{e0}$ for $i = 1,2,\ldots,k$. Therefore, $z-d_e \in H_1$, and, since $-d \in H_1^{*w}$,

$$1 = \|z\|_w^2 = \|d_e\|_w^2+\|z-d_e\|_w^2 + 2(d_e,z-d_e)_w \geq 1+\|z-d_e\|_w^2.$$

Hence, $\|z-d_e\|_w^2 = 0$ or $z = d_e$. The proof is completed.

If the variance σ^2 is unknown, then set $w_i = n_i$ for $i = 0,1,\ldots,k$, let c^* be defined as in (2.5), and reject H_1 for large values of $T_S = T_S'/S$, where S^2 is the usual pooled estimator of σ^2.

3. **COMPARISON OF THE TESTS.** Because of the assumption that $c^* \in H_1^{*w}$, H_0 is least favorable within H_1 for T_S'. Conditioning on S, this is also seen to be the case for its studentized version, T_S. The distributions of the statistics under H_0 and H_1 are either normal or Student's t (noncentral) depending on whether σ^2 is known or not. Hence, tables are available for designing experiments and conducting the test.

At the other extreme, the likelihood ratio tests are extremely complicated. For the case of known variance, the likelihood ratio test rejects H_1 for large values of

$$\bar{\chi}_{12}^2 = \sum_{i=0}^{k} n_i(\mu_i^*-\bar{X}_i)^2/\sigma^2,$$

where $\mu^* = (\mu_0^*,\mu_1^*,\ldots,\mu_k^*)$ is the maximum likelihood estimator of μ restricted by H_1 (see Robertson and Wegman (1978)). A simple computation algorithm for μ^*, which is a consequence of Thompson's algorithm for rooted trees, is discussed in Robertson and Wright (1985). If $\bar{X}_0 \leq \bar{X}_i$ for $i = 1,2,\ldots,k$, then $\mu^* = \bar{X}$. If not, arrange $\bar{X}_1,\bar{X}_2,\ldots,\bar{X}_k$ in increasing order, $\bar{X}_{(1)} \leq \bar{X}_{(2)} \leq \cdots \leq \bar{X}_{(k)}$;

let $n(j)$ denote the sample size corresponding to $\bar{X}_{(j)}$; and determine the smallest positive integer m for which

$$A_m = \left[n_0\bar{X}_0 + \sum_{j=1}^{m} n(j)\bar{X}_{(j)}\right] \Big/ \left[n_0 + \sum_{j=1}^{m} n(j)\right] < \bar{X}_{(m+1)},$$

and if no such integer exists, set $m = k$. Then $\mu_i^* = A_m$ for those i, including 0, with $\bar{X}_i$ included in the computation of A_m, and $\mu_i^* = \bar{X}_i$, otherwise. For the case of unknown variance, the likelihood ratio test rejects H_1 for large values of

$$\bar{E}_{12}^2 = \sigma^2 \bar{\chi}_{12}^2/(\sigma^2 \bar{\chi}_{12}^2 + mS^2),$$

or, equivalently, for large values of $S_{12} = \sigma^2 \bar{\chi}_{12}^2/S^2$. (Note that from the definition of $\bar{\chi}_{12}^2$, $\sigma^2 \bar{\chi}_{12}^2$ does not depend on σ^2, so that $\bar{E}_{12}^2$ and S_{12} are indeed statistics.) The null distributions of $\bar{\chi}_{12}^2$ and $\bar{E}_{12}^2$ are mixtures of chi-squared and beta distributions, respectively. Barlow et al. (1972) give exact formulas for the mixing coefficients for $k \le 3$ and arbitrary sample sizes and table them for $k \le 12$ and equal sample sizes. Robertson and Wright (1985) give approximations for the case $n_0 > n = n_1 = n_2 = \cdots = n_k$ and Wright and Tran (1985) discuss approximations for arbitrary sample sizes.

Because the power function of $\bar{\chi}_{12}^2$ is quite complicated and that of $\bar{E}_{12}^2$ involves a numerical integration of the power of $\bar{\chi}_{12}^2$, we study the powers of these tests for the case in which σ^2 is known. If $k = 2$ and $n_0 = n_1 = n_2$, then, employing the techniques used by Bartholomew (1961), the power of $\bar{\chi}_{12}^2$, i.e., $P[\bar{\chi}_{12}^2 \ge c]$, equals

(3.1) $$\Phi(-\sqrt{c}-\lambda_1)\Phi(\lambda_2)+\Phi(-\sqrt{c}+(\mu_0-\mu_2)/\sqrt{2})\Phi((2\mu_1-\mu_0-\mu_2)/\sqrt{6}) + (2\pi)^{-1}\exp(-\Delta^2/2)\int_{\pi+\beta}^{4\pi/3+\beta}\psi(\Delta\sin\theta,c)d\theta,$$

where $\lambda_1 = (\mu_1-\mu_0)/\sqrt{2}$, $\lambda_2 = (2\mu_2-\mu_1-\mu_0)/\sqrt{6}$, $\Delta\sin\beta = \lambda_1$, $\Delta\cos\beta = \lambda_2$, and

$$\psi(x,c) = (x\Phi(x-\sqrt{c}) + \phi(x-\sqrt{c}))/\phi(x)$$

with Φ and ϕ the cdf and pdf of the standard normal distribution.

For $k = 3$ and $n_0 = n_1 = n_2 = n_3$, we again apply the techniques of Bartholomew (1961). The level sets of μ^* are the sets of indices for which μ_i^* is constant. If μ^* has three level sets, then they may be $\{0,1\},\{2\},\{3\}$; $\{0,2\},\{1\},\{3\}$; or $\{0,3\},\{1\},\{2\}$. We compute the probability that these are the level sets and the test rejects H_1. For the first collection of level sets, the probability is

$$P_3^{(1)} = P[\bar{X}_0 \geq \bar{X}_1,\ \bar{X}_0+\bar{X}_1 < 2\bar{X}_2,\ \bar{X}_0+\bar{X}_1 < 2\bar{X}_3,\ (\bar{X}_1-\bar{X}_0)^2 \geq 2c]$$
$$= P[Z_1 \leq 0,\ Z_2 > 0,\ 2\sqrt{2}\,Z_3+Z_2 > 0,\ Z_1^2 \geq c]$$

with $Z_1 = (\bar{X}_1-\bar{X}_0)/\sqrt{2}$, $Z_2 = (2\bar{X}_2-\bar{X}_0-\bar{X}_1)/\sqrt{6}$ and $Z_3 = (3\bar{X}_3-\bar{X}_0-\bar{X}_1-\bar{X}_2)/\sqrt{12}$. (It should be noted that with $Z_4 = (\bar{X}_0+\bar{X}_1+\bar{X}_2+\bar{X}_3)/2$ this is an orthogonal transformation of $\bar{X}$.) The probability $P_3^{(1)}$ can be rewritten as

(3.2) $$\Phi(-\sqrt{c}-\lambda_1)\int_{-\lambda_2}^{\infty}\Phi(\lambda_3+(\lambda_2+x)/\sqrt{8})\phi(x)dx$$

with λ_1 and λ_2 defined as before and $\lambda_3 = (3\mu_3-\mu_0-\mu_1-\mu_2)/\sqrt{12}$. With μ_1 and μ_2 interchanged, let $P_3^{(2)}$ denote (3.2); let $P_3^{(3)}$ denote (3.2) for the mean vector $(\mu_0,\mu_3,\mu_2,\mu_1)$; and also set $P_3 = P_3^{(1)} + P_3^{(2)} + P_3^{(3)}$.

If μ^* has two level sets, then they may be $\{0,1,2\},\{3\}$; $\{0,1,3\},\{2\}$; or $\{0,2,3\},\{1\}$. The probability that the first collection of level sets occurs and the test rejects H_1 is

$$P_2^{(1)} = P[\bar{X}_0 \geq (\bar{X}_0+\bar{X}_1+\bar{X}_3)/3,\ (\bar{X}_0+\bar{X}_1)/2 \geq (\bar{X}_0+\bar{X}_1+\bar{X}_2)/3,\ (\bar{X}_0+\bar{X}_2)/2$$
$$\geq (\bar{X}_0+\bar{X}_1+\bar{X}_2)/3 < \bar{X}_3,\ \sum_{i=0}^{2} (\bar{X}_i-(\bar{X}_0+\bar{X}_1+\bar{X}_2)/3)^2 \geq c]$$
$$= P[Z_3 > 0]P[Z_2 < 0,\ \sqrt{3}Z_1+Z_2 < 0,\ Z_2-\sqrt{3}Z_1 > 0,\ Z_1^2+Z_2^2 \geq c],$$

where $\{Z_1,Z_2,Z_3\}$ are i.i.d. with the standard normal distribution. Changing to polar coordinates, $Z_1 = R\cos\theta$ and $Z_2 = R\sin\theta$, $P_2^{(1)}$ becomes

$$\Phi(\lambda_3)P[\sin\theta < 0,\ \sin(\theta+\pi/3) < 0,\ \sin(\theta-\pi/3) > 0,\ R^2 \geq c] \tag{3.3}$$
$$= (2\pi)^{-1}\Phi(\lambda_3)\ \exp(-(\lambda_1^2+\lambda_2^2)/2)\int_{\pi}^{4\pi/3} \psi(\Lambda_1,c)d\theta$$

with $\Lambda_1 = \lambda_1\cos\theta + \lambda_2\sin\theta$, and $\lambda_1,\lambda_2,\lambda_3$ and ψ as before. Let $P_2^{(2)}$ be (3.3) for $\mu = (\mu_0,\mu_1,\mu_3,\mu_2)$ and $P_2^{(3)}$ be (3.3) for $\mu = (\mu_0,\mu_3,\mu_2,\mu_1)$ and set $P_2 = P_2^{(1)} + P_2^{(2)} + P_2^{(3)}$.

If μ^* has one level set then, as was observed in the last section, $\bar{X} \in H_1^* \oplus H_0$ or $\bar{X}_i \leq (\bar{X}_0+\bar{X}_1+\bar{X}_2+\bar{X}_3)/4$ for $1 \leq i \leq 3$. Changing to spherical coordinates, $Z_1 = R\cos\beta$, $Z_2 = R\sin\beta\sin\theta$ and $Z_3 = R\sin\beta\cos\theta$ with $0 \leq \beta \leq \pi$, $0 \leq \theta \leq 2\pi$, the probability that μ^* has one level set and $\bar{\chi}^2_{12}$ rejects H_1 is (note $\sin\beta > 0$ a.s.)

$$P_1 = P[\cos \le 0,\ \cos\theta+\sqrt{2}\sin\theta \ge \sqrt{6}\,\mathrm{ctn}\,\beta,\ \cos\theta-2\sqrt{2}\sin\theta > 0,\ R^2 \ge c]$$

$$= P[\pi/2 \le \theta \le 3\pi/2,\ \sin(\theta+\alpha) \ge \sqrt{2}\,\mathrm{ctn}\,\beta,\ \sin(\theta-\gamma) < 0,\ R^2 \ge c],$$

where $\alpha = \sin^{-1}(1/\sqrt{3})$ and $\gamma = \sin^{-1}(1/3)$. Because of the restrictions on θ, $\sin(\theta+\gamma) < 0$, and so we may rewrite P_1 as

$$(2\pi)^{-3/2}\exp(-\Delta^2/2)\int_{\pi+\gamma}^{3\pi/2}\int_{\pi-\tan^{-1}(\sqrt{2}/|\sin(\theta+\alpha)|)}^{\pi}\xi(\Lambda,c)\sin(\beta)\,d\beta d\theta$$

with $\Lambda = \lambda_1\cos\beta+\lambda_2\sin\beta\sin\theta+\lambda_3\sin\beta\cos\theta$, $\Delta^2 = \lambda_1^2+\lambda_2^2+\lambda_3^2$, and $\xi(\Lambda,c) = [(\sqrt{c}+\Lambda)\phi(c-\Lambda)+(1+\Lambda)^2\Phi(\Lambda-\sqrt{c})]/\phi(\Lambda)$. Thus, the power of $\bar{\chi}^2_{12}$ is $P_1+P_2+P_3$.

While the tests based on T_S' are much easier to use than the likelihood ratio tests, we will see that their powers drop off much more rapidly than those of $\bar{\chi}^2_{12}$ or $\bar{E}^2_{12}$ as the alternative moves away from the "center" of the alternative region. We consider multiple contrast tests as compromise procedures. Because orthogonal contrast tests perform very well for testing H_0 versus H_1-H_0 (cf. Mukerjee, Robertson and Wright (1985)), they could be considered in this setting. Suppose $c^{(1)}$ and $c^{(2)}$ are orthogonal, i.e., $(c^{(1)},c^{(2)})_w = \sum_{i=0}^{k} w_i c_i^{(1)} c_i^{(2)} = 0$. We will show that if $x,y \in H_1^{*w}$, $x \ne 0$ and $y \ne 0$, then $(x,y)_w > 0$. Recall, because $x \in H_1^{*w}$, $\sum_{i=0}^{k} w_i x_i = 0$ and $x_i \le 0$ for $i = 1,2,\ldots,k$. Therefore,

$$(x,y)_w = w_0x_0y_0 + \sum_{i=1}^{k} w_ix_iy_i > 0.$$

Because contrasts which are identically zero are not of interest, we suppose $c^{(1)} \neq 0$ and $c^{(2)} \neq 0$. Therefore, $c^{(1)} \notin H_1^{*w}$ or $c^{(2)} \notin H_1^{*w}$, and without loss of generality we assume $c^{(1)} \notin H_1^{*w}$. The significance level for the test which rejects for $\max\{T_{c^{(1)}}, T_{c^{(2)}}\} \geq t$ is no smaller than $\sup_{H_1} P[T_{c^{(1)}} \geq t]$, which, in Section 2, was shown to be one since $c^{(1)} \notin H_1^{*w}$. So reasonable orthogonal contrast tests do not exist in this setting; H_1^{*w} is "too narrow."

Since reasonable orthogonal contrast tests do not exist, we consider tests based on the maximum of several contrasts with the angle between the contrast vectors less than 90°. The distribution theory for such a test is much more manageable if each pair of contrast vectors has the same angle between them. If the treatment sample sizes are equal, i.e., $n = n_1 = n_2 = \cdots = n_k$, the corners for the dual cone H_1^{*w} are $c^{(i)} = (c_{i0}, c_{i1}, \ldots, c_{ik})$ with $n_0c_{i0} = nc_{ii} = (1/n_0+1/n)^{-1}$ and $c_{ij} = 0$ for $j \notin \{0,i\}$. Also, for $i \neq i'$, $(c^{(i)}, c^{(i')})_w = (1+n_0/n)^{-1}$ which corresponds to a 60° angle if $n_0 = n$. Mukerjee, Robertson and Wright (1985) studied the behavior of multiple contrast tests with a common angle between the contrast vectors. They found that for smaller angles the power functions of the tests are quite peaked at the center of the contrast vectors and that they drop off rapidly as the alternative moves away from the center. Based on these findings, we recommend as large an angle as possible, and in particular that the corners of H_1^{*w} be used. This leads to the test statistics

$$T_D' = \max_{1 \leq i \leq k}(\bar{X}_0-\bar{X}_1)/[\sigma(1/n_0+1/n)^{1/2}],$$

and its studentized version, T_D, is given in (1.1). If $n = n_1 = \cdots = n_k$, then $\{(\bar{X}_0-\bar{X}_i)/[S(1/n_0+1/n)^{1/2}] : 1 \le i \le k\}$ has a multivariate t distribution with common correlation $\rho = (1+n_0/n)^{-1}$ (Johnson and Kotz (1976), p. 134). We restrict our attention to this case. Tables of critical values are available for various values of ρ and $m = (n_0-1)+k(n-1)$; see Krishnaiah and Armitage (1965, 1966). If the number of degrees of freedom is large enough that the power of T_D' provides a reasonable approximation to the power of T_D, then the following formula may be used:

(3.4) $$P[T_D' \ge t] = 1-\int_{-\infty}^{\infty}\{\prod_{i=1}^{k} \Phi[t(n/n_0+1)^{1/2}+ \sqrt{n}(\mu_i-\mu_0)-(n/n_0)^{1/2}x]\}\phi(x)dx.$$

While computation of (3.4) involves a numerical integration, the power function of T_D is much more tractable than that of $\bar{\chi}^2_{12}$. In fact, very little is known about the latter for $k > 3$.

Because of the simplicity in using T_S and T_D, it is of interest to compare their power functions with that of the likelihood ratio test. We compare $\bar{\chi}^2_{12}$, T_S' and T_D'. Since all three power functions are invariant under shifts by a constant vector, we may study them in a k-dimensional subspace. For $k = 2$, Bartholomew (1961) considered the parametrization $(\mu_1-\mu_0)/\sqrt{2} = \Delta \sin \beta$ and $(2\mu_2-\mu_1-\mu_0)/\sqrt{6} = \Delta \cos \beta$, with the cone for the simple tree given by $0 \le \beta \le 120°$. For $\Delta = 1,2,3,4$ and $\beta = 120°,130°,\ldots,240°$ the powers of $\bar{\chi}^2_{12}$, T_S' and T_D' are given in Table 1. (Because the power functions are symmetric about $\beta = 240°$, larger β need not be considered.) The following observations should be made from Table 1:

(1) All three tests are biased.

Table 1. Powers of $\bar{\chi}^2_{12}$, T'_s and T'_D with $k = 2$ and $\alpha = .05$.

	$\Delta = 1$			$\Delta = 2$		
β—	$\bar{\chi}^2_{12}$	T'_S	T'_D	$\bar{\chi}^2_{12}$	T'_S	T'_D
120°	.0278	.0160	.0293	.0254	.0041	.0276
130°	.0407	.0235	.0427	.0542	.0099	.0581
140°	.0579	.0345	.0603	.1023	.0232	.1085
150°	.0791	.0500	.0819	.1708	.0500	.1793
160°	.1037	.0706	.1067	.2536	.0972	.2639
170°	.1302	.0963	.1329	.3403	.1683	.3510
180°	.1568	.1261	.1586	.4201	.2595	.4295
190°	.1816	.1581	.1821	.4858	.3597	.4921
200°	.2032	.1897	.2020	.5350	.4551	.5367
210°	.2206	.2180	.2174	.5688	.5347	.5648
220°	.2332	.2403	.2282	.5900	.5927	.5802
230°	.2407	.2546	.2345	.6013	.6273	.5872
240°	.2433	.2595	.2365	.6049	.6387	.5891
	$\Delta = 3$			$\Delta = 4$		
β	$\bar{\chi}^2_{12}$	T'_S	T'_D	$\bar{\chi}^2_{12}$	T'_S	T'_D
120°	.0253	.0008	.0275	.0253	.0001	.0275
130°	.0758	.0038	.0810	.1037	.0013	.1103
140°	.1764	.0152	.1859	.2786	.0097	.2907
150°	.3246	.0500	.3375	.5179	.0500	.5322
160°	.4895	.1306	.5037	.7310	.1711	.7428
170°	.6348	.2681	.6478	.8663	.3911	.8739
180°	.7416	.4425	.7519	.9345	,6388	,9389
190°	.8105	.6116	.8175	,9648	.8229	.9673
200°	.8512	.7432	.8543	.9778	.9221	.9789
210°	.8736	.8298	.8722	.9832	.9656	.9831
220°	.8852	.8798	.8791	.9855	.9827	.9842
230°	.8904	.9048	.8806	.9864	.9891	.9840
240°	.8920	.9123	.8806	.9866	.9907	.9838

(2) All three power functions seem to take on their maxima near $\beta = 240°$.

(3) While the single contrast test has the greatest power of the three at $\beta = 240°$, it cannot be recommended unless the alternatives can be restricted further. For instance, the percentage gain in power over the other two at $\beta = 240°$ is no more than 10%, but for $\beta = 150°$ and $\Delta = 4$, the power of T'_S is about one-tenth of that of

T'_D and $\bar{\chi}^2_{12}$.

(4) For $\Delta = 1$ and 2, T'_D is at least 97% as efficient as $\bar{\chi}^2_{12}$; for $\Delta = 3$, it is at least 98% as efficient as $\bar{\chi}^2_{12}$; and for $\Delta = 4$, the minimum ratio of the power of T'_D to that of $\bar{\chi}^2_{12}$ is .997. These minima occur at $\beta = 240°$. On the other hand, the reciprocal of this ratio takes on even smaller values. For instance, with $\Delta = 4$, and $\beta = 150°$, these powers are of respectable size and the efficiency of $\bar{\chi}^2_{12}$ with respect to T'_D is 97%.

For $k = 3$, the parametrization $(\mu_1-\mu_0)/\sqrt{2} = \Delta \cos \phi$, $(2\mu_2-\mu_1-\mu_0)/\sqrt{6} = \Delta \sin \phi \sin \theta$ and $(3\mu_3-\mu_2-\mu_1-\mu_0)/\sqrt{12} = \Delta \sin \phi \cos \theta$, $0 \le \theta \le 360°$ and $0 \le \phi \le 180°$, which was used in the derivation of the power function of $\bar{\chi}^2_{12}$, is also employed here. For θ and ϕ which are integral multiples of 15° and correspond to points outside of H_1 and $\Delta = 1,2,3,4$, the powers of the three tests were computed. In this search, the minimum of the ratio of the power of T'_D to the power of $\bar{\chi}^2_{12}$ was found to be .97 for $\Delta = 1$ and 2, .98 for $\Delta = 3$, and .99 for $\Delta = 4$. While these values are too numerous to present, the values of the power functions for patterned alternatives are given in Table 2. For a pattern such as (0,-1,-1,1), the powers are given for constant multiples of the pattern with the multiplier chosen so that $\Delta = 1,2,3,4$. Because of the assumption that $n_1 = n_2 = \cdots = n_k$, the power functions are invariant under permutation of the treatment means μ_1, $\mu_2,\ldots,\mu_k$. (The powers for some patterned alternatives are also given in Table 2 for $k = 2$ and $k = 4$.) As can be seen from this table, the minimum of the efficiency of T'_D with respect to $\bar{\chi}^2_{12}$ seems to occur at the middle of H_1^{*w}, i.e., at $\mu \propto (0,-1,-1,-1)$. It should also be noted that for the patterned alternatives considered, the efficiency of $\bar{\chi}^2_{12}$ relative to T'_D takes on values as small as those for T'_D relative to χ^2_{12}. It

Table 2. Powers* of $\bar{\chi}^2_{12}$, T'_S and T'_D for selected patterns of means and $\alpha = .05$.

Pattern	$\bar{\chi}^2_{12}$	T'_S ($\Delta = 1$)	T'_D	$\bar{\chi}^2_{12}$	T'_S ($\Delta = 2$)	T'_D
			$k = 2$			
0 0-1	.1568	.1261	.1586	.4201	.2595	.4295
0-1-1	.2483	.2595	.2365	.6049	.6387	.5891
0-1 1	.0791	.0500	.0819	.1708	.0500	.1793
			$k = 3^{(2)}$			
0 0-1	.1225	.0948	.1262	.3247	.1640	.3420
0-1-1	.1572	.1428	.1565	.3974	.3120	.3940
0 1-1	.0716	.0500	.0748	.1431	.0500	.1534
-1-1-1	.2390	.2595	.2313	.5950	.6387	.5744
-1-1 1	.0908	.0707	.0925	.1863	.0973	.1889
-1 1 1	.0550	.0344	.0588	.1031	.0231	.1149
			$k = 4^{(2)}$			
0 0 0-1	.1016	.0815	.1088	.2761	.1261	.2911
0 0-1-1	.1244	.1081	.1252	.3078	.2037	.3073
0 0 1-1	.0666	.0500	.0700	.1221	.0500	.1368
0-1-1-1	.1557	.1509	.1548	.3935	.3372	.3784
0-1-1 1	.0821	.0653	.0835	.1576	.0841	.1642
0-1 1 1	.0513	.0377	.0573	.0772	.0279	.1033
-1-1-1-1	.2379	.2595	.2275	.5993	.6387	.5656
-1-1-1 1	.0974	.0815	.0961	.1937	.1261	.1908
-1-1 1 1	.0677	.0500	.0692	.1195	.0500	.1266
-1 1 1 1	.0446	.0290	.0485	.0783	.0160	.0873
		$\Delta = 3$	$k = 2$		$\Delta = 4$	
0 0-1	.7416	.4425	.7519	.9345	.6388	.9389
0-1-1	.8920	.9123	.8806	.9866	.9907	.9838
0-1 1	.3246	.0500	.3375	.5179	.0500	.5322
			$k = 3^{(2)}$			
0 0-1	.6292	.2595	.6530	.8722	.3776	.8863
0-1-1	.6987	.5347	.6912	.9055	.7468	.8944
0 1-1	.2732	.0500	.2924	.4526	.0500	.4781
-1-1-1	.8851	.9123	.8691	.9849	.9907	.9807
-1-1 1	.3365	.1308	.3368	.5211	.1714	.5172
-1 1 1	.1972	.0151	.2177	.3348	.0096	.3617
			$k = 4^{(2)}$			
0 0 0-1	.5564	.1854	.5861	.8283	.2595	.8422
0 0-1-1	.5806	.3371	.5749	.8220	.4952	.8148
0 0 1-1	.2432	.0500	.2622	.4100	.0500	.4398
0-1-1-1	.6852	.5762	.6627	.8943	.7894	.8769
0-1-1 1	.2917	.1067	.2964	.4715	.1334	.4680
0-1 1 1	.1348	.0204	.1915	.2324	.0146	.3223
-1-1-1-1	.8851	.9123	.8624	.9844	.9907	.9788
-1-1-1 1	.3427	.1854	.3329	.5256	.2595	.5060
-1-1 1 1	.2137	.0500	.2218	.3503	.0500	.3508
-1 1 1 1	.1443	.0083	.1653	.2546	.0041	.2815

(2) $\mu_0 = 0$ for the μ's indicated.

*For $k = 4$, the powers of $\bar{\chi}^2_{12}$ were estimated by Monte Carlo (10,000 iterations).

is difficult to choose between the two based on power for $k = 3$.

To obtain an indication of the minimum relative efficiency of T'_D with respect to $\bar{\chi}^2_{12}$ for larger k, $k = 4$ and $k = 9$ were also considered. The powers of $\bar{\chi}^2_{12}$ were estimated in these cases by Monte Carlo with 10,000 iteration (standard error $\leq .005$). Because of the size of the parameter space, attention was restricted to patterned alternatives. For $k = 4$, the minimum of the efficiency of T'_D relative to $\bar{\chi}^2_{12}$ for such alternatives is approximately .95 for $\Delta = 1$ and 2 and .96 for $\Delta = 3$ and 4. On the other hand, the efficiency of $\bar{\chi}^2_{12}$ relative to T'_D drops to as low as 70% for $\mu \propto (0,0,-1,1,1)$.

For $k = 9$, the alternatives were restricted further. Patterned alternatives of the form $(0,0,\ldots,0,-1,-1,\ldots,-1)$ with the last b means equal to -1 for $1 \leq b \leq 9$ were considered; see Table 3. For such alternatives, the estimated minimum of the relative efficiency of T'_D with respect to $\bar{\chi}^2_{12}$ is 93% for $\Delta = 1$, 94% for $\Delta = 2$ and 3, and 96% for $\Delta = 4$. Conversely, the estimated minimum for $\bar{\chi}^2_{12}$ with respect to T'_D is 98% for $\Delta = 1$, 90% for $\Delta = 2$, 87% for $\Delta = 3$, and 92% for $\Delta = 4$.

In conclusion, because of its ease of implementation, its less complex power function, and its power relative to the likelihood ratio test, the multiple contrast test based on T'_D should be considered a serious competitor to the likelihood ratio test in this special case. It is anticipated that a comparison of its studentized version, T_D, with the likelihood ratio procedure $\bar{E}^2_{12}$, would result in similar conclusions. However, the even simpler procedures based on T_S cannot be recommended if one cannot restrict the alternatives further than $\mu \notin H_1$.

Table 3. Powers* of $\bar{\chi}^2_{12}$, T'_S, and T'_S with $k = 9$ and $\alpha = .05$.

Pattern	$\Delta = 1$ $\bar{\chi}^2_{12}$	T'_S	T'_D	$\Delta = 2$ $\bar{\chi}^2_{12}$	T'_S	T'_S
b = 1	.0781	.0625	.0794	.1735	.0774	.1936
b = 2	.0852	.0697	.0824	.1720	.0984	.1764
b = 3	.0907	.0768	.0872	.1810	.1134	.1798
b = 4	.0969	.0849	.0932	.1949	.1355	.1911
b = 5	.1064	.0948	.1008	.2170	.1640	.2090
b = 6	.1177	.1081	.1110	.2469	.2037	.2361
b = 7	.1359	.1280	.1263	.2944	.2654	.2794
b = 8	.1651	.1640	.1534	.3829	.3776	.3591
b = 9	.2413	.2595	.2235	.5932	.6387	.5562
		$\Delta = 3$			$\Delta = 4$	
b = 1	.3773	.0948	.4330	.6582	.1150	.7144
b = 2	.3370	.1261	.3575	.5701	.1640	.5927
b = 3	.3389	.1610	.3449	.5545	.2200	.5577
b = 4	.3571	.2037	.3551	.5679	.2890	.5601
b = 5	.3899	.2595	.3812	.6029	.3776	.5874
b = 6	.4430	.3371	.4256	.6651	.4952	.6387
b = 7	.5255	.4532	.4980	.7477	.6524	.7201
b = 8	.6599	.6387	.6233	.8659	.8465	.8407
b = 9	.8825	.9123	.8547	.9827	.9907	.9764

*For $k = 9$, the powers of $\bar{\chi}^2_{12}$ were estimated by Monte Carlo (10,000 iterations).

APPENDIX

The proofs of the two lemmas of Section 2 are given here.

PROOF OF LEMMA 2.3.

Let $v_i = (v_{i0}, v_{i1}, \ldots, v_{ik})$ with $v_{i0} = 1/w_0$, $v_{ii} = -1/w_0$, and $v_{ij} = 0$ for $j \notin \{0, i\}$, $i = 1, 2, \ldots, k$. It is easy to see that $H_1^{*w} = \left\{ \sum_{i=1}^{k} b_i v_i : b_i \ge 0,\ 1 \ge i \ge k \right\}$. Now, $(v_i, E_w(\mu|H_1))_w \le 0$ for all μ and $i = 1, 2, \ldots, k$. Let $c = \sum_{i=1}^{k} b_i v_i$. Since $c \ne 0$, $b_j > 0$ for some j, say j_0. Let $i_0 \in \{1, 2, \ldots, k\} - \{j_0\}$ and define the vector d by $d_{i_0} = -1$ and $d_i = 0$, otherwise. Using either (2.2) or the algorithm discussed in Robertson and Wright (1985) for computing projections for the simple tree, we see that $E_w(d|H_1)_i = -w_{i_0}/(w_0+w_{i_0})$ for $i \in \{0,i_0\}$ and 0 otherwise. Hence, $(v_{j_0},E_w(d|H_1))_w < 0$, and so $(c,E_w(d|H_1))_w \le b_{j_0}(v_{j_0},E_w(d|H_1))_w < 0$. Multiplying d by the appropriate positive constant we obtain the desired μ.

PROOF OF LEMMA 2.4.

As we have seen, $\mu_b - E_w(\mu_b|H_1) = E_w(\mu_b|H_1^{*w}) = E_w(\mu+bE_w(\mu|H_1)|H_1^{*w}) = E_w((b+1)\mu - bE_w(\mu|H_1^{*w})|H_1^{*w}) = E_w(d\mu+(1-d)E_w(\mu|H_1^{*w})|H_1^{*w})$ with $d = b+1$. Applying Theorem 7.5 of Barlow et al. (1972), we see that the last expression is $E_w(\mu|H_1^{*w}) = \mu - E_w(\mu|H_1)$ The proof is completed.

REFERENCES

Abelson, R.P. & Tukey, J.W. (1963). Efficient Utilization of Non-numeric Information in Quantitative Analysis: General Theory and the Case of Simple Order. *Ann. Math. Statist.* **34**, 1347-1369.

Barlow, R.E., Bartholomew, D.J., Bremner, J.M., & Brunk, H.D. (1972) Statistical Inference Under Order Restrictions. New York: Wiley.

Bartholomew, D.J. (1959). A Test of Homogeneity for Ordered Alternatives I, II. *Biometrika* **46**, 36-48 and 328-335.

Bartholomew, D.J. (1961). A Test for Homogeneity of Means Under Restricted Alternatives (with discussion). *J. Roy. Statist. Soc.* **B23**, 239-281.

Brunk, H.D. (1955). Maximum Likelihood Estimates of Monotone Parameters. *Ann. Math. Statist.* **26**, 607-616.

Dunnett, C.W. (1955). A Multiple Comparisons Procedure for Comparing Several Treatments with a Control. *J. Amer. Statist. Assn.* **50**, 1096-1121.

Dykstra, R.L. (1984). Dual Convex Cones of Order Restrictions with Applications. Inequalities in Statistics and Probability. Hayward, CA: Institute of Mathematical Statistics.

Johnson, N.L. & Kotz, S. (1976). Distributions in Statistics: Continuous Multivariate Distributions. New York: Wiley.

Krishnaiah, P.R. & Armitage, P.V. (1965). Percentage Points of the Multivariate T-distribution. Report ARL-654-199, Aerospace Research Laboratories, Wright Patterson Air Force Base, Ohio.

Krishnaiah, P.R. & Armitage, P.V. (1966). Tables for the Multivariate T-distribution. *Sankhya* **B28**, 31-56.

Mukerjee, H., Robertson, T. & Wright, F.T. (1985). Comparison of Several Treatment Means with a Control Using Orthogonal Contrasts. Tech. Report No. 111, Department of Statistics and Actuarial Science, University of Iowa.

Robertson, T. & Wegman E.J. (1978). Likelihood Ratio Tests for Order Restrictions in Exponential Families. *Ann. Statist.* **6**, 485-505.

Robertson, T. & Wright, F.T. (1985). One-sided Comparisons for a Treatment with a Control. *Canad. J. Statist.* **13**, 109-122.

Schaafsma, W. & Smid, L.J. (1966). Most Stringent Somewhere Most Powerful Tests Against Alternatives by a Number of Linear Inequalities. *Ann. Math. Statist.* **37**, 1161-1172.

Wright, F.T. & Tran, T.A. (1985). Approximating the Level Probabilities in Order Restricted Inference: The Simple Tree Ordering. *Biometrika* **72**, 429-439.

Hari Mukerjee
Division of Statistics
University of California
at Davis
Davis, CA 95616

Tim Robertson
Department of Statistics and
Actuarial Science
Iowa City, IA 52242

F. T. Wright
Department of Mathematics
& Statistics
University of Missouri
Rolla, MO 65401

ON TESTING SYMMETRY AND UNIMODALITY[1]

Tim Robertson
The University of Iowa

AMS 1980 subject classifications: Primary 62F03; Secondary 62F04.
Key words and phrases: order restricted tests, isotonic inference, likelihood ratio tests, Chi-bar-square distribution.

ABSTRACT

Likelihood ratio tests of the null hypothesis that a parameter set is both unimodal and symmetric against the alternatives that it is unimodal but not symmetric and vice versa are studied. The parameter set could be multinomial, a set of normal means, or parameters indexing members of an exponential family. Null hypothesis distributions are found for these tests.

1. INTRODUCTION: The distributional properties of symmetry and unimodality play important roles in statistical theory and practice. In practice, particular parametric distributions are fitted to data because they are symmetric (cf. Spiegelhalter (1983)) or because they are unimodal (cf. Bliss and Fisher (1953)) or, implicitly, because they are both. Unimodality is an important example of the type of restriction studied in order restricted inference and the assumption of symmetry allows one to discuss, unequivocally, the location or middle of a distribution and is assumed in many statistical

---

1. This research was sponsored by the Office of Naval Research under ONR contract N00014-80-C321.

investigations. A number of tests for symmetry have been proposed (cf. Lockhart and McLaren (1985), Boos (1982), Antille et al. (1982) and the references therein) and tests for unimodality can be found in Barlow, Bartholomew, Bremner, and Brunk (1972). The two assumptions of symmetry and unimodality are usually investigated or tested independently but often when one or the other of these assumptions is made it is clear that the investigator has both in mind (cf. Antille et al. (1982)).

Both of these concepts can be thought of as order restrictions and they both have to do with the shape of the distribution or of the parameter set. It is the purpose of this note to consider likelihood ratio tests of the null hypothesis that a set of parameters is both unimodal and symmetric when the alternative hypothesis imposes one or the other of these restrictions. Perhaps the most interesting setting for our results is the multinomial and these problems are discussed in Sections Two and Three. In both of these sections we assume that we have in hand the relative frequencies in n independent trials of mutually exclusive events. It will be convenient to think of the vector of parameters and estimates of this vector both as points in a Euclidean space and as functions on the set of integers indexing these parameters. We define the set of symmetric points to be $S = \{x = (x_1, x_2, \cdots, x_k);\ x_i = x_{k-i+1};\ i = 1, 2, \cdots, k\}$ and the set of unimodal points to be $U = \{x;\ x_1 \le x_2 \le \cdots \le x_{\frac{k+1}{2}} \ge \cdots \ge x_k\}$. This is the definition for k odd. For k even we let $U = \{x;\ x_1 \le x_2 \le \cdots \le x_{k/2}, x_{k/2+1} \ge \cdots \ge x_k$. This is a particular type of unimodality with mode in the center. We consider the likelihood ratio tests of H_0 against H_a and against H_a' where

$$
\begin{aligned}
&H_0 : p \in S \cdot U \\
(1.1) \qquad &H_a : p \in U-S \\
&H_a' : p \in S-U
\end{aligned}
$$

where $p = (p_1, p_2, \cdots, p_k)$ is the vector of parameters. The least favorable, asymptotic, null hypothesis distributions for these tests are given in Theorems 2.3 and 2.4.

Since all three of the hypotheses $p \in S \cdot U$, $p \in S$, $p \in U$ can be thought of as imposing quasi-order restrictions on the components of p, the work of Robertson (1978) gives the asymptotic null hypothesis distributions of the likelihood ratio tests of these hypotheses as null hypotheses. The null hypothesis distribution of the test of $p \in S$ is a standard chi-square distribution with $[k/2]$ degrees of freedom. The null hypothesis distribution of the other two test statistics involves the so-called level probabilities of order restricted inference. These probabilities depend upon the particular order restriction under test. For testing $p \in S \cdot U$ they are expressed in terms of a linear order and have been widely investigated (cf. the discussion preceding Theorem 2.4). On the other hand, very little is known about these level probabilities for testing $p \in U$ as a null hypothesis.

In Section Three we consider a contingency table setting and the parameter vector p is assumed to be k^2 dimensional $(p = (p_{ij})_{i,j=1}^{k})$. The set of symmetric vectors is defined to be $S' = \{x = (x_{ij});\ x_{ij} = x_{k-j+1,k-i+1}\}$ and the set of unimodal vectors to be $U' = \{x;\ x_{ij} \leq x_{i',j'}$ if $i \leq i'$, $j \leq j'$, $i+j \leq k+1$, $i'+j' \leq k+1$ and $x_{ij} \geq x_{i',j'}$ if $i \leq i'$, $j \leq j'$, $i+j \geq k+1$, $i'+j' \geq k+1\}$. Think of the vector, x, as a $k \times k$ matrix with i indexing the row and j indexing the column. The restriction that $x \in S'$ requires x to be symmetric around the diagonal running from lower left to upper right. The restriction that $x \in U'$ requires that x is nondecreasing as we move down or to the right and we are above that diagonal and nonincreasing on the other side. We consider the likelihood ratio tests of H_0 against H_a and against H_a' where

$$(1.2) \qquad \begin{aligned} H_0 &: p \in S' \cdot U' \\ H_a &: p \in U' - S' \\ H_a' &: p \in S' - U'. \end{aligned}$$

The least favorable, asymptotic, null hypothesis distributions of these tests are given in Theorems 3.1 and 3.2. In Section Four we discuss the extension of the results developed in Sections Two and Three to other parametric settings.

2. TESTING UNIMODALITY AND SYMMETRY. Assume that k is odd and consider the likelihood ratio test (LRT) of H_0 against H_a where H_0 and H_a are given by (1.1). The likelihood ratio is defined in terms of the maximum likelihood estimates (MLE's) under the restrictions $p \in S \cdot U$ and $p \in U$ so that the first problem is to find these estimates. Both of these constraints are equivalent to requiring that p, as a function on $\{1, 2, \cdots, k\}$, is isotonic with respect to a quasi-order on $\{1, 2, \cdots, k\}$. Thus, by Example 2.1 of Barlow et al. (1972) the maximum likelihood estimates of p subject to these constraints are the appropriate least squares projections of the vector, $\hat{p}$, of relative frequencies onto the given set. For example, the MLE of p subject to $p \in S \cdot U$ is the solution to

$$\min_{q \in S \cdot U} \sum_{i=1}^{k} (\hat{p}_i - q_i)^2 w_i .$$

We denote these projections by $E_w(\hat{p}|S\cdot U)$ and $E_w(\hat{p}|U)$. In case $w_1 = w_2 = \cdots = w_k = 1$, we drop the subscript w and denote the projections by $E(\hat{p}|S\cdot U)$ and $E(\hat{p}|U)$.

For our purposes, the most convenient way to think of computing these projections is via the maximum upper sets algorithm (cf. Barlow et al., pp. 76, 77). The upper sets are subsets of $\{1,2,\cdots,k\}$ and for $p \in U$ they are the sets of consecutive integers containing $\frac{k+1}{2}$. For $p \in S\cdot U$ they are the symmetric sets of consecutive integers containing $\frac{k+1}{2}$, namely, $\left\{\frac{k+1}{2}\right\}, \left\{\frac{k-1}{2}, \frac{k+1}{2}, \frac{k+3}{2}\right\}, \cdots$. In order to compute $E_w(\hat{p}|S\cdot U)$ and $E_w(\hat{p}|U)$, find the upper set which gives the largest average value of $\hat{p}$ over upper sets. The values of the projection for subscripts in this solution set are just the average over the solution set. Eliminate the first solution set from $\{1,2,\cdots,k\}$ and repeat the process on the reduced set, which will consist of the disjoint union of two tails of $\{1,2,,\cdots,k\}$. The pool adjacent violators algorithm, discussed in Section 1.2 of Barlow et al. (1972), can be used to compute the projections on these two tails. We will also think of calculating $E(\hat{p}|S\cdot U)$ iteratively via the following theorem.

THEOREM 2.1. *If* L *is a linear subspace of* R^k *and if* A *is a closed convex cone in* R^k *and if for each* $x \in R^k$, $E(E(x|L)|A) \in L\cdot A$ *then* $E(x|L\cdot A) = E(E(x|L)|A)$.

PROOF. It is straightforward to verify that $E(E(x|L)|A)$ has the properties characterizing $E(x|L \cdot A)$ in Theorem 7.8 of Barlow et al. (1972).

Now S is a linear subspace and U is a closed convex cone and it is clear from the maximum upper sets algorithm that $E(E(\hat{p}|S)|U) \in S \cdot U$ so that $E(\hat{p}|S \cdot U)$ can be computed using Theorem 2.1. If we define x' by $x_i' = x_{k-i+1}$ then it is easy to see that $E(x|S) = \frac{x+x'}{2}$ so that $E(x|S \cdot U) = E\left[\frac{x+x'}{2} |U\right]$. Note that $E(x|S \cdot U)$ can also be computed by defining the $\frac{k+1}{2}$ dimensional vector y by $y_i = \frac{x_i + x_{k-i+1}}{2}$; applying the pool adjacent violators algorithm to y with weights $(2,2,\cdots,2,1)$; and finally setting $E(x|S \cdot U)_i = y_i$; $i = 1,2,\cdots,\frac{k+1}{2}$ and $E(x|S \cdot U)_i = E(x|S \cdot U)_{k-i+1}$ for $i > \frac{k+1}{2}$. Incidentally, it is easy to construct examples to show that $E(x|S \cdot U) \neq E(E(x|U)|S)$.

Now, let Λ be the likelihood ratio for testing H_0 against H_a and let $T = -2 \ln \Lambda$. A likelihood ratio test rejects for large values of T and the asymptotic, null hypothesis, distribution of T can be found using methods similar to those given in the proof of Theorem 2 of Robertson (1978). The following is a sketch of this derivation.

Writing T as a sum and expanding $\ln E(\hat{p}|S \cdot U)_i$ and $\ln E(\hat{p}|U)_i$ about $\ln \hat{p}_i$ using a second degree Taylor's expansion, we obtain

$$T = \sum_{i=1}^{k} \left\{ \frac{n\hat{p}_i}{\alpha_i^2} \left[\hat{p}_i - E(\hat{p}|S \cdot U)_i\right]^2 - \frac{n\hat{p}_i}{\beta_i^2} \left[\hat{p}_i - E(\hat{p}|U)_i\right]^2 \right\}$$

where α_i is between $E(\hat{p}|S \cdot U)_i$ and $\hat{p}_i$, and β_i is between $E(\hat{p}|U)_i$ and $\hat{p}_i$. Assume that p satisfies H_0 and let $U(p)$ be the set of all points in R^k which satisfy the restrictions in U over indices where values of p are equal. For example, if $k = 7$

and $p_1 = p_2 < p_3 = p_4 = p_5 > p_6 = p_7$ then $U(p) = \{x;\ x_1 \le x_2,\ x_3 \le x_4 \ge x_5,\ x_6 \ge x_7\}$. In this example, for sufficiently large n, with probability one $\hat{p}_1 \vee \hat{p}_2 < \hat{p}_3 \wedge \hat{p}_4 \wedge \hat{p}_5 > \hat{p}_6 \vee \hat{p}_7$ so that $E(\hat{p}|S\cdot U) = E(\hat{p}|S\cdot U(p))$ and $E(\hat{p}|U) = E(\hat{p}|U(p))$. This happens in general and moreover, with regard to $E(\cdot|S\cdot U(p))$ and $E(\cdot|U(p))$, p acts as a nonnegative constant so that it can be taken inside the projection operator both as a term and as a factor. Thus, for sufficiently large n with probability one,

$$T = \sum_{i=1}^{k} \left\{\frac{\hat{p}_i}{\alpha_i^2}\left[E(\sqrt{n}(\hat{p}-p)|S\cdot U(p))_i - \sqrt{n}(\hat{p}_i - p_i)\right]^2 \right.$$
$$\left. - \frac{\hat{p}_i}{\beta_i^2}\left[E((\sqrt{n}(\hat{p}-p)|U(p))_i - \sqrt{n}(\hat{p}_i - p_i)\right]^2\right\}.$$

Now $\hat{p}_i$, $E(p|S\cdot U)_i$, $E(\hat{p}|U)_i$ and therefore α_i and β_i converge almost surely to p_i and the vector $\sqrt{n}(\hat{p}-p)$ converges in distribution to $Y = (Y_1, Y_2, \cdots, Y_k)$ where $Y_i = p_i(V_i - \bar{V})$, $\bar{V} = \sum_{i=1}^{k} p_i V_i$, and $V_1, V_2, \cdots, V_k$ are independent and $V_i \sim n(0, p_i^{-1})$. Thus, T converges in distribution to

$$\sum_{i=1}^{k} \left\{\frac{1}{p_i}\left[E(Y|S\cdot U(p))_i - Y_i\right]^2 - \frac{1}{p_i}\left[E(Y|U(p))_i - Y_i\right]^2\right\}$$
$$= \sum_{i=1}^{k} p_i\left[E(V|S\cdot U(p))_i - V_i\right]^2 - p_i\left[E(V|U(p))_i - V_i\right]^2\Big\}$$
$$= \sum_{i=1}^{k} \left\{\left[E(Z|S\cdot U(p))_i - Z_i\right]^2 - \left[E(Z|U(p))_i - Z_i\right]^2\right\}$$

where $Z_1, Z_2, \cdots, Z_k$ are independent standard normal variables.

These limiting distributions depend upon p through U(p) and our task, now, is to find the largest, stochastically, of them. The following theorem provides the necessary tool. Let $\langle x,y \rangle_w$ denote the inner product, $\langle x,y \rangle_w = \sum_{i=1}^{k} x_i y_i w_i$, and let $\|x\|_w = \langle x,x \rangle_w^{1/2}$ with the convention that we drop the subscript w when $w_i = 1;\ i = 1,2,\cdots,k$.

THEOREM 2.2. *Assume that* $A \subset R^k$ *is a closed convex cone having the properties that* $\|x-E_w(x|A)\|_w^2 = \|x'-E_w(x'|A)\|^2$ *for all* $x \in R^k$ $(x_i' = x_{k-i+1})$ *and* $E_w(E_w(X|S)|A) \in S \cdot A$ *for all* $x \in R^k$. *If* $L(x) = \|x-E_w(X|S \cdot A)\|_w^2 - \|x-E_w(X|A)\|_w^2$ *then* $L(x) \le \|x-E_w(x|S)\|_w^2$ *for all* $x \in R^k$.

PROOF. $L(x) = \|x-E_w(x|S)\|_w^2 + 2\langle x-E_w(x|S), E_w(x|S)-E_w(x|S\cdot A)\rangle_w$

$$+ \|E_w(x|S)-E_w(x|S\cdot A)\|_w^2 - \|x-E_w(x|A)\|_w^2.$$

The inner product is zero since S is a linear subspace so that it suffices to argue that $\|E_w(x|S)-E_w(x|S\cdot A)\|_w^2 \le \|x-E_w(x|A)\|_w^2$. However,

$$\begin{aligned}\|E_w(x|S)-E_w(x|S\cdot A)\|_w^2 &= (1/4)\|x+x'-E_w(x+x'|A)\|_w^2 \\ &\le (1/4)\|x+x'-(E_w(x|A)+E_w(x'|A))\|_w^2 \\ &\le 1/4\{2[\|x-E_w(x|A)\|_w^2 + \|x'-E_w(x'|A)\|_w^2]\} \\ &= \|x-E_w(x|A)\|_w^2.\end{aligned}$$

The first inequality is because $E_w(x|A) + E_w(x'|A) \in A$ and the second follows from (7.1) on page 314 of Barlow et al. (1972).

Now, in the limiting distribution of T, $U(p)$ is a closed convex cone satisfying the assumption placed on A in Theorem 2.2. Thus for any real number t and any $p \in S \cdot U$,

$$\lim_{n\to\infty} P_p[T \geq t] \leq P[\|Z-E(Z|S)\|^2 \geq t].$$

Moreover, if we are dealing with a p such that $p_1 < p_2 < \cdots p_{\frac{k+1}{2}} > p_{\frac{k+3}{2}} > \cdots > p_k$ then $U(p) = R^k$ so that $P[\|Z-E(Z|S)\|^2 \geq t]$ is actually the supremum. Since $E(Z|S) = \frac{Z+Z'}{2}$ it is straightforward to show that $\|Z-E(Z|S)\|^2$ is the sum of $\frac{k-1}{2}$ squares of independent standard normal variables. A similar result holds when k is even.

THEOREM 2.3. *If* $T = -2 \ln \Lambda$ *where* Λ *is the likelihood ratio for testing* H_0 *against* H_a *then*

$$\sup_{p\in H_0} \lim_{n\to\infty} P_p[T \geq t] = P[\chi^2_{\frac{k-1}{2}} \geq t]$$

if k *is odd, and*

$$\sup_{p\in H_0} \lim_{n\to\infty} P_p[T \geq t] = P[\chi^2_{\frac{k}{2}} \geq t]$$

if k *is even.*

Now consider testing the null hypothesis, H_0, against the alternative $H_a' : p \in S-U$. Since both the null and alternative hypotheses impose the symmetry restriction on p, this problem is equivalent to the problem obtained by amalgamating the events indexed by i and by $k-i+1$ and then testing the null hypothesis that the vector q defined by $q_i = \frac{p_i+p_{k-i+1}}{2}$; $i = 1,2,\cdots,\frac{k+1}{2}$ has non-decreasing components. The maximum likelihood estimate of q can be obtained using the pool adjacent violators algorithm on $\hat{q}$ where $\hat{q}_i = \frac{\hat{p}_i+\hat{p}_{k-i+1}}{2}$; $i = 1,2,\cdots,\frac{k+1}{2}$ and the null hypothesis distribution of the likelhood ratio statistic $T^* = -2 \ln \Lambda$ is given by Theorem 2 of Robertson (1978). We obtain the following results.

THEOREM 2.4. *Let* $T^* = -2 \ln \Lambda$ *where* Λ *is the likelihood ratio for testing* H_0 *against* H_a' *given by* (1.1). *Then*

$$\sup_{p \in H_0} \lim_{n\to\infty} P_p[T^* \geq t] = \sum_{\ell=1}^{H} P_w(\ell,H)P[\chi^2_{H-\ell} \geq t]$$

where $P_w(\ell,H)$ *is the probability that* $E_w(W|L)$ *has exactly* ℓ *distinct values*, $H = \frac{k+1}{2}$, $w = (2,2,\cdots,2,1)$ *if* k *is odd and* $H = \frac{k}{2}$, $w = (2,2,\cdots,2)$ *if* k *is even.*

Computation of the probabilities $P_w(\ell,H)$, is discussed in Barlow et al. (1972), Section 3.3 and an approximation for the unequal weights case is given in Robertson and Wright (1983). A computer routine to implement this approximation is given in Pillers et al. (1984).

3. **CONTINGENCY TABLES.** Let $\hat{p}_{ij}$; $i,j = 1,2,\cdots,k$ be the relative frequencies in n independent trials of k^2 mutually exclusive events with probabilities p_{ij}; $i,j = 1,2,\cdots,k$. Let $R = -2 \ln \Lambda$ where Λ is the likelihood ratio for testing H_0 against H_a and

H_0 and H_a are given in (1.2). The maximum likelihood estimates subject to $p \in S' \cdot U'$ and $p \in U'$ are again least squares projections, $E(\hat{p}|S' \cdot U')$ and $E(\hat{p}|U')$, from Example 2.1 in Barlow et al. (1972). Also by the maximum upper sets algorithm for $E(\cdot|U')$, $E(E(x|S')|U') \in S' \cdot U'$ so by Theorem 2.1, $E(x|S' \cdot U') = E(E(x|S')|U')$ and it is clear that $E(x|S') = \frac{x+x'}{2}$ where $x'_{ij} = x_{k-j+1,k-i+1}$; $i,j = 1,2,\cdots,k$.

The techniques developed in Section 2 carry over to this problem so that $R \xrightarrow{\mathcal{L}} \|Z - E(Z|S' \cdot U'(p))\|^2 - \|Z - E(Z|U'(p))\|^2$ where $U'(p)$ imposes the constraints in U' only over subscripts where p is constant and we are assuming that $p \in S' \cdot U'$. Theorem 2.2 carries over to this new setting with $x'_{ij} = x_{k-j+1,k-i+1}$ so that $\lim_{n\to\infty} P_p[R \geq t] \leq P[\|Z - E(Z|S')\|^2 \geq t]$ with Z_{ij}; $i,j = 1,2,\cdots,k$ independent standard normal variables. Moreover, for any $p \in H_0$ for which none of the values p_{ij} on the same side of the diagonal are equal, $U'(p) = R^{k^2}$ and we obtain the following result.

THEOREM 3.1. *Let* $R = -2 \ln \Lambda$ *where* Λ *is the likelihood ratio for testing* H_0 *against* H_a *given in* (1.2). *Then*

$$\sup_{p\in H_0} \lim_{n\to\infty} P_p[T \geq t] = P[\|Z - E(Z|S')\|^2 \geq t]$$
$$= P\left[\chi^2_{\frac{k^2-k}{2}} \geq t\right].$$

Now consider the likelihood ratio test of H_0 against H'_a where H_0 and H'_a are given by (1.2). As with Theorem 2.4 the problem can be handled by amalgamating the events indexed by (i,j) and (k-j+1, k-i+1) and applying standard techniques. We obtain the following result.

THEOREM 3.2. *If* $R^* = -2 \ln \Lambda$ *where* Λ *is the likelihood ratio for testing* H_0 *against* H_a' *where* H_0 *and* H_a' *are given by* (1.2) *then*

$$\sup_{p \in H_0} \lim_{n\to\infty} P[R^* \geq t] = \sum_{\ell=1}^{\frac{k^2+k}{2}} P_w(\ell,k) P\left[\chi^2_{\frac{k^2+k}{2}-\ell} \geq t\right]$$

where $P_w(\ell,k)$ *is the probability that* $E_w(W|L)$ *has exactly* ℓ *distinct values.*

No good algorithms for the computation of $P_w(\ell,k)$ exist at the present time. If the value of k is moderate these could be computed via a Monte Carlo simulation. If this is done the most efficient algorithm for computing $E_w(W|L)$ is given in Dykstra and Robertson (1982). A computer program to implement this algorithm is given in Bril et al. (1984).

4. TESTS FOR SYMMETRY AND UNIMODALITY UNDER OTHER POPULATION ASSUMPTIONS. Suppose we have independent random samples from each of k normal populations with means $\mu_1, \mu_2, \cdots, \mu_k$ and we wish to test $H_0 : \mu \in S \cdot U$ against $H_a : \mu \in U - S$ where U and S are the subsets of R^k defined as in (1.1). Assume first that the population variances, $\sigma_1^2, \sigma_2^2, \cdots, \sigma_k^2$, are known and let the sample size and sample means be denoted by $n_1, n_2, \cdots, n_k$ and $\bar{X}_1, \bar{X}_2, \cdots, \bar{X}_k$. The maximum likelihood estimates of μ subject to $\mu \in S \cdot U$ and $\mu \in U$

are given by $E_w(\overline{X}|S\cdot U)$ and $E_w(\overline{X}|U)$ where $w_i = n_i/\sigma_i^2$. These projections are computed as in Section Two.

If $T = -2 \ln \Lambda$ where Λ is the likelihood ratio then $T = \sum_{i=1}^{k} \{w_i[\overline{X}_i - E_w(\overline{X}|S\cdot U)_i]^2 - w_i[\overline{X}_i - E_w(\overline{X}|U)_i]^2\}$. If the weights are symmetric (i.e., $w_i = w_{k-i+1}$) then we can apply Theorem 2.2 to obtain

$$\begin{aligned} P_\mu[T \ge t] &\le P_\mu\left[\sum_{i=1}^{k} w_i[\overline{X}_i - E(\overline{X}|S)_i]^2 \ge t\right] \\ &= P_0\left[\sum_{i=1}^{k} w_i[\overline{X}_i + \mu_i - E(\overline{X}+\mu|S)_i]^2 \ge t\right] \\ &= P_0\left[\sum_{i=1}^{k} w_i[\overline{X}_i - E(\overline{X}|S)_i]^2 \ge t\right] \end{aligned}$$

since if $\mu \in S\cdot U$ then $E(\overline{X}+\mu|S) = E(\overline{X}|S)+\mu$. Moreover if k is odd and we take $\mu = (\delta, 2\delta, \cdots, \frac{k+1}{2}\delta, \frac{k-1}{2}\delta, \cdots, \delta)$ and let $\delta \to \infty$ we find that $P_\mu[T > t]$ converges to this upper bound. We have the following result.

THEOREM 4.1. *Under the assumptions described above,*

$$\begin{aligned} \sup_{\mu \in S\cdot U} P_\mu[T \ge t] &= P\left[\chi^2_{\frac{k+1}{2}} \ge t\right] \quad \textit{if } k \textit{ is odd} \\ &= P\left[\chi^2_{\frac{k}{2}} \ge t\right] \quad \textit{if } k \textit{ is even.} \end{aligned}$$

Theorem 4.1 is analogous to Theorem 2.4 of Warrack and Robertson (1984) and has some negative implications concerning the likelihood ratio tests of a null hypothesis of symmetry and unimodality against an alternative of unimodality but not symmetry. Let $T' = -2 \ln \Lambda$ where Λ is the likelihood ratio for testing the null hypothesis that $\mu \in S$ against all alternatives $\left[T' = \sum_{i=1}^{k} w_i[\bar{X}_i - E(\bar{X}|S)_i]^2\right]$ and consider using T' to test H_0 against H_a. Then for any $\mu \in S \cdot U$, $P_\mu[T' \geq t] = P\left[\chi^2_{\frac{k-1}{2}} \geq t\right]$ so that T and T' would have the same critical values and moreover by Theorem 2.2, $T \leq T'$ so that T' is uniformly more powerful than T for testing H_0 against H_a. This is a rather surprising result since a test based upon T' neglects much information about μ, namely that $\mu \in U$. It is a consequence of the least favorable configuration for T $(\lim_{\delta\to\infty}(\delta, 2\delta, \cdots, \delta))$ and the fact that the power function of T' is constant over H_0. Intuition tells us that we should be able to do better than T' but the author doesn't know how to accomplish this. Perhaps the critical value for T should be adjusted to account for the fact that in most problems μ is bounded. An alternative to the LRT might be to do a conditional test; an idea which is just now beginning to be researched (cf. Wollan and Dykstra (1986)).

In testing $\mu \in S \cdot U$ against $\mu \in S - U$ we do not need to assume that the weights are symmetric. The proof of the following result uses the same techniques as the proof of Theorem 2.4.

THEOREM 4.2. *If* $T = -2 \ln \Lambda$ *where* Λ *is the likelihood ratio for testing* $H_0 : \mu \in S \cdot U$ *against* $H_a' : \mu \in S - U$ *then*

$$\sup_{\mu \in S \cdot U} P_\mu[T \geq t] = \sum_{\ell=1}^{k} P_{w^*}(\ell, H) P[\chi^2_{H-\ell} \geq t]$$

where $w_i^* = w_i + w_{k-i+1}$; $i = 1,2,\cdots,H$ *and* $H = k/2$ *if* k *is even* *and* $w_i^* = w_i + w_{k-i+1}$; $i = 1,2,\cdots,H-1$, $w_H^* = w_H$ *and* $H = \frac{k+1}{2}$ *if* k *is odd.*

Let $T' = -2 \ln \Lambda$ where Λ is the likelihood ratio for testing the null hypothesis that $\mu \in U$ against all alternatives (i.e., $T' = \sum_{i=1}^{k} w_i[\overline{X}_i - E(\overline{X}|U)_i]^2$). It is not true that the power of T' dominates the power of T in Theorem 4.2. Using a Monte Carlo simulation for $k = 6$ we compared the power of T' to that of T at 241 points in S-U. The power of T was larger than that of T' at each of these points and often by as much as 35%. The author is not able to explain why the likelihood ratio gives rise to an apparently reasonable test in the testing situation described in Theorem 4.2 and fails in the testing situation described in Theorem 4.1.

Now consider the case where the population variances are unknown but assumed to be equal and let Λ be the likelihood ratio for testing $H_0 : \mu \in S \cdot U$ against the alternative $H_a : \mu \in U-S$. The likelihood ratio can be written $\Lambda = (\hat{\sigma}_1^2/\hat{\sigma}_0^2)^{N/2}$ where $\hat{\sigma}_0^2$ and $\hat{\sigma}_1^2$ are the maximum likelihood estimates of the common value, σ^2, of the population variances under $\mu \in S \cdot U$ and $\mu \in U$. Specifically,

$$\hat{\sigma}_0^2 = N^{-1} \sum_{i=1}^{k} \sum_{j=1}^{n_i} (X_{ij} - E_w(\overline{X}|S \cdot U)_i)^2 \qquad (w_i = n_i)$$

$$\hat{\sigma}_1^2 = N^{-1} \sum_{i=1}^{k} \sum_{j=1}^{n_i} (X_{ij} - E_w(\overline{X}|U)_i)^2.$$

A likelihood ratio test rejects for large values of $S = 1-\Lambda^{2/N}$. Adding and subtracting $\bar{X}_i$ in both $\hat{\sigma}_0^2$ and $\hat{\sigma}_1^2$, expanding the squares, and a little algebra together with the observation that the inner product terms are zero yields

$$S = \frac{\|\bar{X}-E_w(\bar{X}|S\cdot U)\|_w^2-\|\bar{X}-E_w(\bar{X}|U)\|_w^2}{\|\bar{X}-E_w(\bar{X}|S\cdot U)\|_w+\Sigma_{i=1}^k \ \Sigma_{j=1}^{n_i}(X_{ij}-\bar{X}_i)^2}$$

$$= \frac{\|\bar{X}-E_w(\bar{X}|S\cdot U)\|_w-\|\bar{X}-E_w(\bar{X}|U)\|_w^2}{\|\bar{X}-E_w(\bar{X}|S)\|_w^2 + \|E(\bar{X}|S)-E(\bar{X}|S\cdot U)\| +\Sigma_{i=1}^k\Sigma_{j=1}^n(X_{ij}-\bar{X}_i)^2}$$

$$\leq \frac{\|\bar{X}-E_w(\bar{X}|S)\|_w^2}{\|\bar{X}-E_w(\bar{X}|S)\|_w^2+\Sigma_{i=1}^k\Sigma_{j=1}^{n_i}(X_{ij}-\bar{X}_i)^2},$$

assuming that $w_i = w_{k-i+1}$ so that Theorem 2.2 can be applied to the numerator and by dropping a nonnegative term in the denominator. If, in addition, we assume that $\mu \in S\cdot U$ then $\|\bar{X}-E(\bar{X}|S)\|_w^2/\sigma^2$ has a chi-square distribution with H degrees freedom ($H = \frac{k-1}{2}$ if k is odd and $H = \frac{k}{2}$ if k is even). $\sum_{i=1}^{k}\sum_{j=1}^{n_i}(X_{ij}-\bar{X}_i)^2/\sigma^2$ has a chi-square distribution with $N-k$ degrees freedom and these two random variables are independent. Thus, for any t and $\mu \in S\cdot U$, $P_\mu[S \geq t] \leq P\left[B_{\frac{H}{2},\frac{N-k}{2}} \geq t\right]$ where $B_{\alpha,\beta}$ is a beta variable with parameters α and β. Moreover, by letting $\mu = (\delta, 2\delta, \cdots, \delta)$ and $\delta \longrightarrow \infty$ as in Theorem 4.2 we can show that this gives the supremum. We have the following theorem.

THEOREM 4.3. *If* $S = 1-\Lambda^{2/N}$ *where* Λ *is the likelihood ratio for testing* H_0 *against* H_a, *then when* $\sigma_1^2 = \sigma_2^2 = \cdots = \sigma_k^2$,

$$\sup_{\mu \in H_0} P_\mu[S \geq t] = P\left[B_{\frac{H}{2},\frac{N-k}{2}} \geq t\right]$$

where $B_{\alpha,\beta}$ *is a standard beta variable with parameters* α *and* β.

The discussion following Theorem 4.1 also applies to Theorem 4.3.

It is clear that the techniques developed in this paper can be applied to yield null hypothesis distributions for likelihood ratio statistics in a number of testing problems. As in Robertson and Wegman (1978), samples from exponential populations could be considered. Theorems analogous to Theorems 3.1 and 3.2 can be proven for likelihood ratio tests regarding a matrix $\mu = (\mu_{ij})_{i,j=1,2,\cdots,k}$ of normal means. Finally we could find null hypothesis distributions in these testing problems under the assumption that the samples were taken from populations having densities which belong to a regular exponential family as in Robertson and Wegman (1978). Of course assumptions on the sample sizes and auxiliary parameters need to be made in order to yield symmetric weights so that Theorem 2.2 can be applied.

Most of the tests which have been studied for testing symmetry involve sampling from a continuous population. The tests discussed in Sections 2 and 3 could be used in these sampling situations if one could decide how to "discretize" the continuous problem (i.e., pick "cells" with probabilities $p_1, p_2, \cdots, p_k$). This does not seem to be an easy question and is the subject of current research efforts. It appears at the present time that the alternative should play an important role in the choice of these "cells."

REFERENCES

Antille, A., Kersting, G., and Zucchini, W. (1982). Testing symmetry. *J. Amer. Statist. Assoc.* **77**, 639-646.

Barlow, R.E., Bartholomew, D.J., Bremner, J.M., and Brunk, H.D. (1972). *Statistical Inference under Order Restrictions*, John Wiley & Sons.

Bliss, C.I. and Fisher, R.A. (1953). Fitting the negative binomial distribution to biological data and a note on the efficient fitting of the negative binomial. *Biometrics* **9**, 176-200.

Boos, D.D. (1982). A test for asymmetry associated with the Hodges-Lehman estimator. *J. Amer. Statist. Assoc.* **77**, 647-651.

Bril, G., Dykstra, R.L., Pillers, C., and Robertson, Tim (1984). A Fortran program for isotonic regression in two independent variables, *Applied Statistics* **33**, 352-357.

Dykstra, R.L. and Robertson, Tim (1982). An algorithm for isotonic regression for two or more independent variables, *Ann. Statist.* **10**, 708-716.

Lockhart, R.A. and McLaren, C.G. (1985). Asymptotic points for a test of symmetry about a specified median. *Biometrika* **72**, 208-210.

Pillers, C., Robertson, Tim, and Wright, F.T. (1984). A Fortran program for the level probabilities of order restricted inference, *Applied Statistics* **33**, 115-119.

Robertson, Tim (1978). Testing for and against an order restriction on multinomial parameters, *J. Amer. Statist. Assoc.* **73**, 197-202.

Robertson, Tim and Wegman, Edward J. (1978). Likelihood ratio tests for order restrictions in exponential families. *Ann. Statist.* **6**, 485-505.

Robertson, Tim and Wright, F.T. (1983). On approximation of the level probabilities and associated distributions in order restricted inference. *Biometrika* **70**, 597-606.

Spiegelhalter, D.J. (1983). Diagnostic tests of distributional shape. *Biometrika* **70**, 401-409.

Warrack, G. and Robertson, Tim (1984). A likelihood ratio test regarding two nested but oblique order restricted hypotheses, *J. Amer. Statist. Assoc.* **79**. 881-886.

Wollan, P.C. and Dykstra, R.L. (1986). Conditional tests with an order restriction as a null hypothesis. *Advances in Order Restricted Statistical Inference*, Springer-Verlag.

Tim Robertson
Department of Statistics & Actuarial Science
University of Iowa
Iowa City, IA 52242

ON TESTS UNDER ORDER RESTRICTIONS IN REDUCTION OF DIMENSIONALITY[1]

Ashis SenGupta
University of Wisconsin
and
Indian Statistical Institute

AMS 1980 subject classificatons: *Primary 62H15; Secondary 62F05.*
Keywords and phrases: *generalized canonical variables, generalized variance, isotonic regression.*

ABSTRACT

In techniques for reduction of dimensionality, initially the components of the original vector variable are grouped into several disjoint subsets. New variables with a reduced dimension are then constructed. Often several meaningful alternative groupings can be formed. Optimal choice of the grouping and/or the dimension for the new variables is of considerable importance. Using a generalization (SenGupta, 1983) of canonical variables, this leads naturally to tests under order restrictions for the generalized variances of the generalized canonical variables. By suitable transformations, it is seen that a solution can be given by an appeal to isotonic regression.

1. INTRODUCTION. Canonical variable analysis (Anderson, 1984) plays an important role in applied research since it provides a method of meaningfully creating bivariate random variables from two groups,

---

1. Research supported in part by NSF Grant SE 579-13976 and ONR Contract N00014-75-C-0442.

each consisting possibly of a large number of components. For the case of more than two groups, SenGupta (1982) constructed generalized canonical variables (GCV's) which possess some desirable properties both for theoretical and practical purposes. For a brief review, see SenGupta (1983). These GCV's are equicorrelated and their generalized variance (GV) is minimal.

It is assumed that a meaningful grouping of $\underline{X} : p \times 1$ into several disjoint subvectors $\underline{X}_1, \cdots, \underline{X}_k, \underline{X}_j : p_j \times 1$, $\sum_{j=1}^{k} p_j = p$ is already given. Let us denote the first GCV by $\underline{Y}$. It may be useful to determine whether with the same dimension k, reasonable alternative regroupings of $\underline{X}$ produce better results (see, e.g., Gnanadesikan, 1977, p. 77). If there are s reasonable groupings, each of dimension k, then one will test, H_0 : the $|\Sigma_{{}_i\underline{Y}}|$ are all equal against H_1 : the $|\Sigma_{{}_i\underline{Y}}|$ are in a given order, where ${}_i\underline{Y} : k \times 1$ is the first GCV, $\mathrm{Disp}({}_i\underline{Y}) = \Sigma_{{}_i\underline{Y}}$, for the ith grouping, $i = 1, \cdots, s$. (Obviously, GCV's with smaller GV's are to be preferred.) Since the GCV's are equicorrelated, the above hypotheses can also be formulated in terms of the equicorrelation coefficients, ρ_i, where the ρ_i are the common correlation coefficients between the components of ${}_i\underline{Y}$. Let $\underline{X}$ be distributed as $N_p(\underline{O}, \Sigma)$. Then ${}_i\underline{Y}$ is distributed exactly or asymptotically (SenGupta, 1982) as $N_k(\underline{O}, \Sigma_{\rho_i})$ according as Σ is known or unknown, where Σ_{ρ_i} is a correlation matrix with all off-diagonal elements equal to ρ_i. We show that the likelihoood ratio test for H_0 against H_1 can be derived by an appeal to the technique of isotonic regression.

2. DEFINITIONS AND PRELIMINARY RESULTS. We recall the definition of the new GCV's as in SenGupta (1982).

DEFINITION 1. The first new GCV is the vector $\underline{Y}^{(1)\prime} = [\underline{Y}^{(1)}, \cdots, Y_k^{(1)}] = [\underline{\alpha}'_{(1)}\underline{X}^{(1)}, \cdots, \underline{\alpha}'_{(k)}X^{(k)}]$, where $\underline{\alpha}_i^{(1)}$, $i = 1, \cdots, k$ are chosen such that the $Y_i^{(1)}$, $i = 1, \cdots, k$ have unit variances, are equicorrelated with equicorrelation coefficient, say ρ, and the generalized variance of $\underline{Y}^{(1)}$, $|\Sigma_\rho|$, is minimum. $\underline{\alpha}$ and ρ are solutions to (SenGupta, 1983)

$$\begin{bmatrix} -(k-1)\rho\Sigma_{11} & \Sigma_{12} & \cdots & \Sigma_{1k} \\ \Sigma_{21} & -(k-1)\rho\Sigma_{22} & \cdots & \\ \Sigma_{k1} & \Sigma_{k2} & \cdots & -(k-1)\rho\Sigma_{kk} \end{bmatrix} \begin{bmatrix} \underline{\alpha}_1 \\ \vdots \\ \underline{\alpha}_k \end{bmatrix} = 0,$$

i.e., $(\Sigma - \lambda^* \Sigma_d)\underline{\alpha} = \underline{0}$, where $\Sigma = \text{Disp}(\underline{X})$, $\lambda^* = 1 + (k-1)\rho\alpha\Sigma_d$ is a diagonal super matrix with block diagonal matrices Σ_{11}, $i = 1, \cdots, k$.

The criterion of GV has been considered, among others, by Anderson (1984), Kettenring (1971) and Steel (1951). We have imposed the constraint of equicorrelation since it guarantees the validity of the usual test procedures in MANOVA, growth curve analysis, etc., using the GCV's even when $n < k$.

Higher stage, e.g., second, third, etc., new GCV's can be constructed as in SenGupta (1982). However, for our testing purposes we will restrict ourselves to $\underline{Y}^{(1)}$ to be denoted by $\underline{Y}$ henceforth, and to the corresponding $\text{Disp}(\underline{Y}) = \Sigma_\rho$.

If Σ is known, then $\underline{\alpha}$ is known. However, if Σ is unknown, then Σ and also $\underline{\alpha}$ are to be estimated. In the former case, $\underline{Y}$ is distributed exactly as $N_k(0, \Sigma_\rho)$ while in the latter case the result is only approximate in large samples.

Suppose a random sample of size n, $\underline{X}_1,\cdots,\underline{X}_n$ is taken from $N_p(0,\Sigma)$. Let $\underline{Y}$ and $\underline{Z}$ be first new GCV's corresponding to two modes of grouping. Construct $\underline{Y}_i = (\underline{\alpha}'_{Y_1}\underline{X}^*_{1i},\cdots,\underline{\alpha}'_{Y_k}\underline{X}^*_{ki})$ and $Z_i = (\alpha'_{\underline{Z}_1}\underline{X}^{**}_{1i},\cdots,\alpha'_{\underline{Z}_k}\underline{X}^{**}_{ki})'$ for each $\underline{X}_i$, $i = 1,\cdots,n$, where $\underline{X}' = (X^{*\prime}_1,\cdots,\underline{X}^{*\prime}_k)$ and $X'_- = (\underline{X}^{**\prime}_1,\cdots,\underline{X}^{**\prime}_k)$ are the partitions of $\underline{X}$ into k sets corresponding to two different groupings. Then $(\underline{Y}_1,\cdots,\underline{Y}_n)$ and $(\underline{Z}_1,\cdots,\underline{Z}_n)$ can each be considered separately as a sample of GCV's corresponding to $\underline{Y}$ and $\underline{Z}$, respectively. But $\underline{Y}_i$ and $\underline{Z}_j$ will still be dependent for $i = j$. This dependency can be avoided if two independent subsamples of $\underline{X}$ are drawn and one is used to construct $\underline{Y}$ and the other to construct $\underline{Z}$. But then a much reduced sample size for the GCV's as compared to n will be obtained. Alternatively, we use a transformation as shown below, on the n GCV's obtained from our original sample $\underline{X}_1,\cdots,\underline{X}_n$, of size n, to give uncorrelated variables. These will be used for testing purposes.

LEMMA 1. *If* $\underline{X}_1,\cdots,\underline{X}_n$ *are i.i.d. and* $\underline{Y}_1,\cdots,\underline{Y}_n$ *are i.i.d.,* $\underline{X}_i$ *is independent of* $\underline{Y}_j$, $i \neq j$, *and the* $\underline{X}_i$*'s and the* $\underline{Y}_i$*'s have multivariate normal distributions, then* $\underline{X} = \Sigma a_i\underline{X}_i$ *and* $\underline{Y} = \Sigma b_i\underline{Y}_i$ *are independent if* $\Sigma a_i b_i = 0$.

PROOF.
$$\begin{aligned}\mathrm{Cov}(\underline{X},\underline{Y}) &= \Sigma\Sigma a_i b_j \,\mathrm{Cov}(\underline{X}_i,\underline{Y}_j) = \Sigma a_i b_i \,\mathrm{Cov}(\underline{X}_i,\underline{Y}_i)\\ &= \mathrm{Cov}(\underline{X}_1,\underline{Y}_1)\cdot 0 = 0,\end{aligned}$$

and hence the lemma follows from the normality of $\underline{X}$ and $\underline{Y}$.

Let $U_j = \sum_{s=1}^{k} Y_{js}$, $Y^* = \sum_{j=1}^{n} a_j U_j$; $V_j = \sum_{s=1}^{k} Z_{js}$, $Z^* = \sum_{j=1}^{n} b_j V_j$.

Choose a_j, b_j such that $\Sigma a_j b_j = 0$. Also if $\underline{Y} \sim N_k(\underline{0}, \Sigma_{\rho_1})$, $\underline{Z} \sim N_k(\underline{0}, \Sigma_{\rho_2})$ then $Var(Y^*) = k\{1+(k-1)\rho_1\}$ and $Var(Z^*) = k\{1+(k-1)\rho_2\}$ where we have chosen a_j, b_j further such that $\Sigma a_j^2 = \Sigma b_j^2 = 1$. Call the resulting variables Y^{**} and Z^{**}.

3. TESTS FOR EQUALITY OF MULTIDIMENSIONAL SCATTER OF SEVERAL NEW GCV's. Consider the case where s (say, even) alternative meaningful groupings of $\underline{X}$ with the same number of subsets k is given, i.e., we have s new GCV's, ${}_1\underline{Y}, \cdots, {}_s\underline{Y}$ each of dimension k. Let $Cov({}_i\underline{Y}) = \Sigma_{\rho_i}$. Then H_0 and H_1 of Section 1 can be restated as $H_0 : |\Sigma_{\rho_i}|$ are all equal against $H_1 : |\Sigma_{\rho_i}|$ are in a specified order, say $|\Sigma_{\rho_i}| > |\Sigma_{\rho_j}|$, $i > j$. Suppose further that ρ_i, $i = 1, \cdots, s$ are all $>$ (or all $<$) 0.

LEMMA 2. *Let* $\rho_i > (<)\ 0$, $i = 1, \cdots, s$. *Then* $H_1 : |\Sigma_{\rho_i}| > |\Sigma_{\rho_j}|$, $i > j$, *is equivalent to* $H_1' : \rho_i < (>)\ \rho_j$, $i > j$.

PROOF. It is straightforward to show that $|\Sigma_\rho| \uparrow$ for $-1/(k-1) < \rho < 0$ and $\downarrow$ for $0 < \rho < 1$, from which the lemma is obvious.

Take $s/2$ independent subsamples, each of size n, from the distribution of $\underline{X}$ and from the ith subsample construct $(y_{2i-1}^{**}, y_{2i}^{**})$, $i = 1, \cdots, s/2$, as in Section 2. An independent estimate of Σ is assumed to be available. Thus the y_i^{**}'s are (approximately) independent normal variables. Then, the likelihood function is given by L, where, writing $y_i = y_i^{**}$,

$$\ln L = C - \frac{1}{2}\,\Sigma\{\ln[k\{1+(k-1)\rho_i\}] + y_i^2/[k\{1+(k-1)\rho_i\}]\}$$

where $C = (2\pi)^{s/2}$. Then, in the notation of Barlow et al. (1972) letting

$$X = \{1,\cdots,s\},\quad g(i) = y_i^2,\quad f(i) = [k\{1+(k-1)\rho_i\}]$$

and $w(i) = 1$, $i = 1,\cdots,s$, we have the

LEMMA 3. *The isotonic regression* g^* *maximizes* L *under* H_1 : *The generalized variance of* i^{th} GVC *is greater than that of the* j^{th} GVC, $i > j$.

PROOF. Note first that since equicorrelations are all < 0 (all > 0), an ordering of the generalized variances of the new GCV's gives the same (reverse) ordering of the corresponding equicorrelation coefficients. With the above values of X, g(i), f(i) and w(i), we use Theorem 1.10 of Barlow et al. (1972) and the relevant notation. Let $\varphi(u) = -\ln u$. To complete the proof it suffices to note that maximizing ln L is equivalent to minimizing $\sum_i [\log f(i)+\{g(i)/f(i)\}]$ both subject to f isotonic, since the first and last terms of Δ_φ do not involve f.

H_0 is equivalent to $H_0' : \rho_1 = \cdots = \rho_k = \rho$ (unknown), say. Then under H_0, the MLE of ρ, say $\hat{\rho}$, is obtained from $k\{1+(k-1)\hat{\rho}\} = \sum_{i=1}^{s} y_i^2/s$.

The likelihood ratio test can now be easily performed. However, even under H_0 the exact distribution of the test statistic is complicated. In large samples, $-2 \ln \lambda \sim \bar{\chi}^2$ (see Barlow et al., 1972, p. 198).

4. EXAMPLES AND DISCUSSIONS. Several authors, including Kettenring, have discussed this example. Three (= k) sets of scores by several people on three batteries of three tests each ($p_1 = 3$, $i = 1,2,3$) were obtained. The three tests in each battery were intended to measure, respectively, the verbal, the numerical and the spatial abilities of the persons tested. The sample correlation matrix of the original scores was given. Gnanadesikan (1977) commented, "An interesting alternative analysis in this example would be to regroup the nine variables into three sets corresponding to the three abilities measured rather than the three batteries of tests." How profitable this alternative grouping would be in terms of multidimensional scatter can be judged by the test for H_0 above. Here $s = 2$, and one would test $H_0 : |\Sigma_{{}_1\underline{Y}}| = |\Sigma_{{}_2\underline{Y}}|$ against $H_1 : |\Sigma_{{}_1\underline{Y}}| < |\Sigma_{{}_2\underline{Y}}|$ where ${}_1\underline{Y}$ and ${}_2\underline{Y}$ correspond to the groupings of Gnanadesikan and Kettenring, respectively. (Unfortunately, the original sample observations seem to be unpublished and hence a detailed numerical example cannot be presented here.)

We have considered above the case where the new GCV's, under alternate regroupings, are of same dimension, k. However, it may be possible to reasonably regroup $\underline{X}$ into various groups of sizes k_i, $i = 1,\cdots,t$, yielding GCV's of different dimensions. It is then worthwhile to explore whether a GCV of smaller dimension performs as well as, if not better than, one of a higher dimension. For such a case, one would test

$$H_0 : |\Sigma_{{}_i\underline{Y}}|^{1/k_i} < |\Sigma_{{}_j\underline{Y}}|^{1/k_j}, \quad k_i < k_j, \quad i \neq j, \quad i,j = 1,\cdots,t,$$

where ${}_j\underline{Y}$ is the first GCV of dimension k_j obtained by the jth mode of regrouping, $j = 1,\cdots,t$.

Though we have considered here new GCV's, similar formulations of testing problems involving order restrictions are valid for GCV's proposed by other authors (e.g., Kettenring, 1971). These seem to be interesting problems for future research.

ACKNOWLEDGEMENTS. The author is grateful to Professors T. W. Anderson, I. Olkin and C. R. Rao for their encouragement and helpful comments.

REFERENCES

Anderson, T.W. (1984). *An Introduction to Multivariate Statistical Analysis*. John Wiley: New York.

Barlow, R.E., Bartholomew, D.J., Bremner, J.M. and Brunk, H.D. (1972). *Statistical Inference Under Order Restrictions*. John Wiley: New York.

Gnanadesikan, R. (1977). *Methods for Statistical Data Analysis of Multivariate Observations*. John Wiley: New York.

Kettenring, J.R. (1971). Canonical analysis of several sets of variables. *Biometrika* **58**, 433-451.

SenGupta, A. (1982). On the problems of construction and statistical inference associated with a generalization of canonical variables. Tech. Rep. 52, Dept. of Statistics, Stanford University.

SenGupta A. (1983). Generalized Canonical Variables, *Encyclopedia of Statistical Sciences*, Vol. 3 (eds. Johnson and Kotz). John Wiley: New York, 123-126.

Steel, R.G.D. (1951). Minimum generalized variance for a set of linear functions. *Ann. Math. Statist.* **22**, 456-460.

Ashis SenGupta
University of Wisconsin-Madison and Indian Statistical Institute
Madison, WI U.S.A. INDIA

POWER SERIES APPROXIMATIONS TO THE NULL DISTRIBUTIONS OF SOME CHI-BAR-SQUARE STATISTICS(1)

Bahadur Singh
Memphis State University

F. T. Wright
University of Missouri-Rolla

AMS 1980 subject classifications: Primary 62F03; Secondary 62E20.

Keywords: approximate significance levels, chi-bar-square distributions, Laguerre polynomials, series approximations, simple tree ordering and total ordering.

ABSTRACT

Some power series approximations to the exact null distribution of the chi-bar-square statistic for several testing situations are developed using the first four cumulants of the null distributions, and their performance is investigated numerically. The series expansions use Laguerre polynomials and the associated gamma densities. Chi-bar-square statistics arise when testing the homogeneity of normal means with the alternative restricted by a partial ordering against all alternatives. Approximations are provided for the case of a total order and a simple tree with equal, or nearly equal, sample sizes. The numerical investigations indicate the accuracy and usefulness of these approximations.

---

1. This research was partially sponsored by the Office of Naval Research under ONR contract N00014-80-C0322.

1. **INTRODUCTION.** We consider situations in which one wishes to test hypotheses about normal means which involve order restrictions. For instance, one may wish to test homogeneity, $H_0: \mu_1 = \mu_2 \cdots = \mu_k$, with the alternative restricted by the total ordering, $H_1: \mu_1 \leq \cdots \leq \mu_k$. On the other hand, one may wish to test H_1 versus $H_2: \mu_1 > \mu_{i+1}$ for some i. In comparing several treatments with a control, a test of H_0 with the alternative restricted by the simple tree ordering $H_1': \mu_1 \leq \mu_i$ for $i = 2,3,\cdots,k$ and of H_1' versus $H_2' : \mu_1 > \mu_i$ for some $i = 2,3,\cdots,k$ are of interest. If the common variance of these normal populations is known, then the likelihood ratio test statistics have null distributions which are mixtures of chi-square distribu- tions, which Bartholomew (1959) called chi-bar-square statistics. They also provide approximations for large degrees of freedom.

The chi-bar-square distributions also arise as approximations when considering multinomial parameters (Robertson (1978)), one-parameter exponential families (Robertson and Wegman (1978)), Poisson intensities (Magel and Wright (1984)) and nonparametric tests (Shirley (1977) and Robertson and Wright (1985)).

A great deal of information (e.g., the locatiOn, variability about the mean, skewness and kurtosis of a distribution) is contained in the first four moments of a distribution, and so we consider four-moment approximations for these chi-bar-square distributions. A natural choice is to use the first four terms of a series expansion involving Laguerre polynomials and the associated gamma distributions. It should be noted that numerical investigations show that using higher moments, such as the fifth and sixth, does not seem to improve the approximation enough to warrant the extra effort. Sasabuchi and Kulatunga (1985) provide similar approximations using the first three moments for the test of H_0 versus H_1-H_0 with unknown variance and they are based on expansions using Jacobi polynomials and the associated beta distributions.

The approximations presented here are based on the first four moments, or equivalently on the first four cumulants, of the chi-bar-square distributions. Because the mixing coefficients for these distributions are intractable for unequal sample sizes and even moderate k, we restrict attention to the case of equal sample sizes. However, Robertson and Wright (1983) and Wright and Tran (1985) have shown that the chi-bar-square distributions are robust to moderate changes in the sample sizes for both the total order and the simple tree. Hence, the approximations should be reasonable if there is not too much variation in the sample sizes.

Approximations for the totally ordered case are presented in Section 2. The simple tree ordering is considered in Section 3 and the results of our numerical investigation are summarized in Section 4. Bartholomew (1959, p. 330) proposed a two-moment approximation which is equivalent to using the first term, i.e., the zero$^{\underline{th}}$ order term, in the Laguerre expansion. The chi-bar-square distributions may assign positive probability to $\{0\}$ and so we show how the two- and four-moment approximations can be corrected for the discrete part. This type of correction was employed by Sasabuchi and Kulatunga (1985). We found that, independent of the value of k, the corrected two-moment approximation is adequate except in the far right tail of the chi-bar-square distributions, but to the right of the $99^{\underline{th}}$ percentile the increase in accuracy warrants the use of the corrected four-moment approximation.

Roy and Tiku (1962), Tiku (1964, 1965, 1971, 1975), Tan and Wong (1977, 1978, 1980), and Hirotsu (1979) ahve used Laguerre series approximations to approximate the sampling distributions of F-ratios in the analysis of variance problems and related topics.

2. SERIES APPROXIMATIONS: THE TOTALLY ORDERED CASE. In this section, we consider approximations to the null distributions of the likelihood ratio test of H_0 versus H_1-H_0 and of H_1 versus H_2 based on Laguerre polynomial expansions.

Assume that $\{y_{ij};\ j = 1,\cdots,n\}$ for $i = 1,\cdots,k$ are independent random samples from k normally distributed populations with mean μ_i and common variance σ^2. Consider the hypotheses H_0, H_1 and H_2 as defined in the introduction, i.e.,

$$H_0 : \mu_1 = \mu_2 = \cdots = \mu_k$$

$$H_1 : \mu_1 \le \mu_2 \le \cdots \le \mu_k$$

$$H_2 : \mu_1 > \mu_{i+1} \quad \text{for some } i.$$

When σ^2 is known, the likelihood ratio test of H_0 versus H_1-H_0 rejects H_0 for large values of

$$T_{01} = n \sum_{i=1}^{k} (\mu_i^* - \hat{\mu})^2/\sigma^2$$

where $\mu^* = (\mu_1^*,\cdots,\mu_k^*)$ is the maximum likelihood estimate of $\mu = (\mu_1,\cdots,\mu_k)$ under H_1 and $\hat{\mu} = \sum_{i=1}^{k}\sum_{j=1}^{n} y_{ij}/nk$. Under H_0,

$$\mathrm{pr}(T_{01} \geq t) = \sum_{\ell=2}^{k} P(\ell,k)\,\mathrm{pr}(\chi^2_{\ell-1} \geq t), \quad t > 0,$$

$$\mathrm{pr}(T_{01} = 0) = 1/k$$

where $P(\ell,k)$ denotes the probability, under H_0, that the coordinates of μ^* have exactly ℓ distinct values, and $\chi^2_{\ell-1}$ denotes a standard chi-squared variable having $\ell-1$ degrees of freedom, with $\chi^2_0 \equiv 0$, cf. Barlow et al. (1972). The likelihood ratio test of H_1 versus H_2 rejects H_1 for large values of

$$T_{12} = n \sum_{i=1}^{k} (\mu^*_i - \bar{y}_i)^2/\sigma^2 \quad \text{with} \quad \bar{y}_i = \sum_{j=1}^{n} y_{ij}/n.$$

H_0 is least favorable within H_1, and under H_0

$$\mathrm{pr}(T_{12} \geq t) = \sum_{\ell=1}^{k-1} P(\ell,k)\,\mathrm{pr}(\chi^2_{k-\ell} \geq t), \quad t > 0,$$

$$\mathrm{pr}(T_{12} = 0) = 1/k!,$$

cf. Robertson and Wegman (1978).

To compute a p-value for either T_{01} or T_{12}, one needs to obtain the $P(\ell,k)$ either from Table A.5 of Barlow et al. (1972) if $k \leq 12$ or from their recursive relation, p. 145, for $k > 12$, and then compute the $k-1$ chi-square tail probabilities. Hence, approximations are of interest for large k.

In the following paragraphs, four approximations to the null distribution of the statistic T_{01} are presented in detail, and the corresponding approximations to the null distribution of the statistic

T_{12} are described very briefly.

2.1 *Approximations to the Null Distribution of* T_{01}. Four series approximations to the null distribution of T_{01} are now discussed.

(i) Four-Moment Approximation of T_{01}.

First, the null distribution of T_{01} is approximated by a scaled gamma density. That is, $T_{01} \approx \rho X_b$ where $\rho > 0$ and X_b has density

$$g_b(x) = \frac{1}{\Gamma(b)} x^{b-1} e^{-x}, \qquad x > 0,$$

i.e., the gamma density with parameters $(b,1)$. Equating the first two cumulants of T_{01} with those of $g_b(x)$, one obtains

$$b = \kappa_1/\rho, \qquad \rho = \kappa_2/\kappa_1 \tag{2.1}$$

where κ_1 and κ_2 are the first two cumulants as in equation (3.47) of Barlow et al. (1972, p. 151). Then, following Davis (1976), Gideon and Gurland (1977), and Kotz, Johnson and Boyd (1967a,b), it can be shown that the probability density function of $X = T_{01}/\rho$ can be expanded in a convergent infinite series involving Laguerre polynomials and the associated gamma densities as

$$\begin{aligned} f(x) &= \left\{1 + \sum_{j=3}^{\infty} c_j L_j^b(x)\right\} g_b(x) \\ &= g_b(x) + \sum_{j=3}^{\infty} d_j \sum_{s=0}^{j} \binom{j}{s} (-1)^s g_{b+s}(x) \end{aligned}$$

where

$$L_j^b(x) = \frac{1}{j!} \sum_{s=0}^{j} \binom{j}{s} (-x)^s \frac{\Gamma(b+j)}{\Gamma(b+s)}$$

is the Laguerre polynomial of degree j, and

$$d_j = c_j \binom{b+j-1}{j} = E\{L_j^b(X)\}.$$

To approximate the distribution, only the terms up to and including $j = 4$ are retained. That is, with $f(x)$ the density of T_{01}/ρ,

$$f(x) \approx g_b(x) + \sum_{j=3}^{4} d_j \sum_{s=0}^{j} \binom{j}{s} (-1)^s g_{b+s}(x) \tag{2.2}$$

where

$$d_3 = \frac{1}{3!}(-\kappa_3^* + 2b)$$

$$d_4 = \frac{1}{4!}(\kappa_4^* - 12\kappa_3^* + 18b) \tag{2.3}$$

$$\kappa_3^* = \kappa_3/\rho^3, \quad \kappa_4^* = \kappa_4/\rho^4$$

and κ_3, κ_4 are the third and fourth cumulants given by equation (3.47) in Barlow et al. (1972, p. 151).

(ii) Four-Moment Approximation of T_{01} with Correction.

Note that $pr(T_{01} = 0) = 1/k$. Therefore, the characteristic function of the conditional distribution of T_{01} given that $T_{01} > 0$, is given by

$$\phi^*(t) = (\phi(t)-k^{-1})/(1-k^{-1})$$

where

$$\phi(t) = (z+1)(z+2)\cdots(z+k-1)/k!$$

and

$$z = (1-2it)^{-1/2}.$$

The first four cumulants of the conditional distribution of T_{01} given that $T_{01} > 0$, are given by

$$\kappa_1^{**} = \frac{k}{(k-1)} \kappa_1$$

$$\kappa_2^{**} = \frac{k}{(k-1)}(\kappa_2+\kappa_1^2) - \frac{k^2}{(k-1)^2} \kappa_1^2$$

$$\kappa_3^{**} = \frac{k}{(k-1)}(\kappa_3+3\kappa_2\kappa_1+\kappa_1^3) - 3\frac{k^2}{(k-1)^2} \kappa_1(\kappa_2+\kappa_1^2) + 2\frac{k^3}{(k-1)^3} \kappa_1^3$$

$$\kappa_4^{**} = \frac{k}{(k-1)}(\kappa_4+3\kappa_2^2+4\kappa_1\kappa_3+6\kappa_1^2\kappa_2+\kappa_1^4) - 4\frac{k^2}{(k-1)^2} \kappa_1(\kappa_3+3\kappa_2\kappa_1+\kappa_1^3)$$

$$- 3 \frac{k^2}{(k-1)^2}(\kappa_2+\kappa_1^2) + 12\frac{k^3}{(k-1)^3} \kappa_1^2(\kappa_2+\kappa_1^2) - 6\frac{k^4}{(k-1)^4} \kappa_1^4$$

where $\kappa_1,\cdots,\kappa_4$ are the cumulants given by equation (3.47) in Barlow et al. (1972, p. 151).

The corrected four-moment approximation is obtained by taking

(2.4)
$$b = \kappa_1^{**}/\rho, \qquad \rho = \kappa_2^{**}/\kappa_1^{**}$$

$$d_3 = \frac{1}{3!} (-\kappa_3^*+2b)$$

$$d_4 = \frac{1}{4!} (\kappa_4^*-12\kappa_3^*+18b)$$

$$\kappa_3^* = \kappa_3^{**}/\rho^3 \quad \text{and} \quad k_4^* = \kappa_4^{**}/\rho^4$$

in the series expansion for $f(x)$ in (2.2).

In particular, let

$$\overline{G}_b(x) = \int_x^\infty g_b(u)du.$$

For $t > 0$, under H_0, $pr(T_{01} \geq t)$ is approximated by

$$(1-k^{-1}) \sum_{j=0}^{4} a_j \overline{G}_{b+j}(t/\rho) \tag{2.5}$$

with

$$a_0 = 1+d_3+d_4,\quad a_1 = -(3d_3+4d_4),\quad a_2 = (3d_3+6d_4),\quad a_3 = -(d_3+4d_4)$$
$$\text{and}\quad a_4 = d_4. \tag{2.6}$$

For $5 \leq k \leq 40$, the values of b, ρ, d_3 and d_4 are given in Table 1.

(iii) Two-Moment Approximation of T_{01}.

In the two-moment approximation, the first two cumulants of the exact null distribution of T_{01} are made equal to those of a scaled gamma distribution, and it can be obtained as a special case of the four-moment series approximation by taking $d_3 = d_4 = 0$ in (2.2). That is, under H_0,

$$pr(T_{01} \geq t) \approx \overline{G}_b(t/\rho) \quad \text{for} \quad t > 0 \tag{2.7}$$

where b and ρ are given by (2.1). Note that this approximation is due to Bartholomew (1959, p. 330).

Table 1. Coefficients for the Corrected Two-Moment and Four-Moment Approximations to the Null Distribution of T_{01}.

k	ρ	b	d_3	d_4
5	2.31791	0.69207	0.01352	0.01691
6	2.37111	0.73383	0.01650	0.02083
7	2.41322	0.77006	0.01908	0.02428
8	2.44757	0.80213	0.02137	0.02735
9	2.47626	0.83093	0.02343	0.03013
10	2.50066	0.85709	0.02529	0.03267
11	2.52174	0.88108	0.02700	0.03501
12	2.54017	0.90325	0.02858	0.03717
13	2.55646	0.92386	0.03004	0.03919
14	2.57098	0.94313	0.03141	0.04108
15	2.58403	0.96122	0.03269	0.04285
16	2.59584	0.97828	0.03390	0.04453
17	2.60658	0.99442	0.03504	0.04612
18	2.61641	1.00973	0.03612	0.04762
19	2.62544	1.02431	0.03714	0.04906
20	2.63378	1.03823	0.03812	0.05043
21	2.64151	1.05153	0.03906	0.05174
22	2.64870	1.06428	0.03995	0.05299
23	2.65541	1.07651	0.04081	0.05420
24	2.66168	1.08828	0.04163	0.05536
25	2.66757	1.09961	0.04242	0.05647
26	2.67310	1.11054	0.04318	0.05755
27	2.67832	1.12110	0.04392	0.05859
28	2.68325	1.13131	0.04463	0.05959
29	2.68791	1.14119	0.04532	0.06056
30	2.69233	1.15077	0.04598	0.06150
31	2.69654	1.16006	0.04662	0.06242
32	2.70053	1.16909	0.04725	0.06330
33	2.70434	1.17785	0.04786	0.06416
34	2.70798	1.18638	0.04845	0.06500
35	2.71145	1.19469	0.04902	0.06581
36	2.71477	1.20277	0.04958	0.06661
37	2.71796	1.21066	0.05012	0.06738
38	2.72101	1.21835	0.05065	0.06813
39	2.72394	1.22586	0.05116	0.06887
40	2.72675	1.23319	0.05167	0.06958

(iv) Two-Moment Approximation of T_{01} with Correction.

The two-moment approximation to the null distribution of T_{01} with correction is obtained by using (2.7) where now b and ρ are given by (2.4). Hence, under H_0,

$$\text{pr}(T_{01} \geq t) = (1-k^{-1})\bar{G}_b(t/\rho) \quad \text{for} \quad t > 0, \tag{2.8}$$

and the values of b and ρ are given in Table 1. This kind of approximation with correction is suggested by Sasabuchi and Kulatunga (1985) in approximating the null distribution of the E-bar-square statistic.

2.2 *Approximations to the Null Distribution of* T_{12}. Note that $\text{pr}(T_{12} = 0) = 1/k!$, which is small even for moderately large values of k, and so, correcting for the discrete part may not improve the approximation significantly. Therefore, only two approximations to the null distribution of T_{12} are given.

The characteristic function of the null distribution of T_{12} is

$$\phi(t) = E(e^{itT_{12}}) = \frac{(z+1)(z+2)\cdots(z+k-1)}{z^{k-1}(k!)}$$

where $z = (1-2it)^{1/2}$. The cumulant generating function is thus

$$\psi(t) = \ln \phi(t) = \sum_{j=1}^{k-1} \ln(z+j) - (k-1)\ln z - \ln k!.$$

The first four cumulants of T_{12} are given by

$$\kappa_1 = (k-1) - \sum_{j=2}^{k} j^{-1}$$

$$\kappa_2 = 2(k-1) - \sum_{j=2}^{k} j^{-1} - \sum_{j=2}^{k} j^{-2}$$

$$\kappa_3 = 8(k-1) - \sum_{j=2}^{k} 3j^{-1} - \sum_{j=2}^{k} 3j^{-2} - \sum_{j=2}^{k} 2j^{-3} \tag{2.9}$$

$$\kappa_4 = 48(k-1) - \sum_{j=2}^{k} 15j^{-1} - \sum_{j=2}^{k} 15j^{-2} - \sum_{j=2}^{k} 12j^{-3} - \sum_{j=2}^{k} 6j^{-4}.$$

(i) Four-Moment Approximation of T_{12}.

Again let

$$b = \kappa_1/\rho \quad \text{and} \quad \rho = \kappa_2/\kappa_1 \tag{2.10}$$

where κ_1 and κ_2 are the first two cumulants of T_{12} given by (2.9). Then, the four-moment approximation to the null distribution of T_{12}/ρ is given by (2.2) and (2.3) where now κ_3 and κ_4 are the third and fourth cumulants of T_{12} given by (2.9). In particular, for $t > 0$, under H_0, $pr(T_{12} \geq t)$ is approximated by

$$\sum_{j=0}^{4} a_j \bar{G}_{b+j}(t/\rho) \tag{2.11}$$

with a_j given by (2.6) and b, ρ, d_3 and d_4 are given in Table 2 for $5 \leq k \leq 40$.

Table 2. Coefficients for the Two-Moment and Four-Moment Approximations to the Null Distribution T_{12}.

k	ρ	b	d_3	d_4
5	2.30174	1.18027	0.03277	0.04535
6	2.27003	1.56386	0.03977	0.05490
7	2.24530	1.96283	0.04618	0.06363
8	2.22537	2.37360	0.05207	0.07164
9	2.20891	2.79370	0.05752	0.07903
10	2.19505	3.22135	0.06258	0.08588
11	2.18319	3.65527	0.06729	0.09225
12	2.17290	4.09443	0.07171	0.09823
13	2.16388	4.53809	0.07587	0.10384
14	2.15589	4.98562	0.07978	0.10911
15	2.14876	5.43651	0.08349	0.11414
16	2.14235	5.89037	0.08700	0.11888
17	2.13656	6.34687	0.09035	0.12339
18	2.13128	6.80571	0.09352	0.12765
19	2.12646	7.26665	0.09656	0.13181
20	2.12203	7.72950	0.09947	0.13567
21	2.11795	8.19407	0.10227	0.13949
22	2.11417	8.66020	0.10501	0.14326
23	2.11066	9.12780	0.10752	0.14656
24	2.10740	9.59670	0.11004	0.14987
25	2.10434	10.06682	0.11247	0.15350
26	2.10148	10.53808	0.11471	0.15623
27	2.09880	11.01036	0.11708	0.15976
28	2.09627	11.48364	0.11914	0.16213
29	2.09389	11.95782	0.12120	0.16499
30	2.09164	12.43284	0.12329	0.16814
31	2.08951	12.90865	0.12532	0.17099
32	2.08749	13.38521	0.12721	0.17388
33	2.08557	13.86249	0.12902	0.17501
34	2.08375	14.34039	0.13097	0.17884
35	2.08201	14.81892	0.13281	0.18221
36	2.08036	15.29805	0.13444	0.18362
37	2.07877	15.77775	0.13608	0.18514
38	2.07727	16.25794	0.13788	0.19053
39	2.07582	16.73867	0.13936	0.18951
40	2.07444	17.21983	0.14125	0.19188

3. SERIES APPROXIMATIONS: THE SIMPLE TREE ORDERING.

In this section, we consider approximations to the null distributions of the likelihood ratio test of H_0 versus $H_1'-H_0$ and of H_1' versus H_2'

based on Laguerre polynomial expansions. Recall, $H_1': \mu_1 \le \mu_i$ for $i = 2,3,\cdots,k$ and $H_2': \mu_1 > \mu_1$ for some $i = 2,3,\cdots,k$. As in Section 2, we let y_{ij}, $1 \le j \le n$ and $1 \le i \le k$, denote the observations with $y_{ij} \sim N(\mu_i,\sigma^2)$ and consider the case of known variances. If $\tilde{\mu} = (\tilde{\mu}_1,\tilde{\mu}_2,\cdots,\tilde{\mu}_k)$ denotes the maximum likelihood estimate of μ subject to the restriction $\tilde{\mu} \in H_1'$, then the likelihood ratio test rejects H_0 for large values of

$$T_{01}' = n \sum_{i=1}^{k} (\tilde{\mu}_i-\hat{\mu})^2/\sigma^2$$

and under H_0,

$$pr(T_{01}' \ge t) = \sum_{\ell=2}^{k} Q(\ell,k)pr(\chi_{\ell-1}^2 \ge t), \quad t > 0$$

$$pr(T_{01}' = 0) = Q(1,k)$$

where $Q(\ell,k)$ is the probability, under H_0, that the coordinates of $\tilde{\mu}$ have exactly ℓ distinct values, cf. Barlow et al. (1972). The likelihood ratio test of H_1' versus H_2' rejects H_1' for large values of

$$T_{12}' = n \sum_{i=1}^{k} (\tilde{\mu}_i-\bar{y}_i)^2/\sigma^2,$$

H_0 is least favorable within H_1' and under H_0,

$$pr(T_{12}' \ge t) = \sum_{\ell=1}^{k-1} Q(\ell,k)pr(\chi_{k-\ell}^2 \ge t), \quad t > 0$$

$$pr(T_{12}' = 0) = Q(k,k)\ 1/k,$$

cf. Robertson and Wegman (1978). For $k \le 12$, the $Q(\ell,k)$ are given in Table A.6 of Barlow et al. (1972) and for $k > 12$, they may be obtained from their (3.38) and (3.39). However, a numerical integration is needed to obtain $Q(\ell,k)$ for $2 \le \ell \le k$.

The characteristic functions of T'_{01} and T'_{12} are given by

$$\phi_1(t) = \sum_{\ell=1}^{k} Q(\ell,k)(1-2it)^{-\frac{\ell-1}{2}} \quad \text{and} \quad \phi_2(t) = \sum_{\ell=1}^{k} Q(\ell,k)(1-2it)^{-\frac{k-\ell}{2}},$$

respectively. Carrying out the numerical integrations needed to compute $Q(\ell,k)$, one can obtain the first four cumulants of T'_{01} and T'_{12}. We see from Table A.6 of Barlow et al. (1972) that $Q(1,k)$ is converging to zero fairly rapidly, i.e., $Q(1,5) < .01$ and hence we need not correct for the discrete part of T'_{01}. As is the case for the approximations without correction, $pr(T'_{01} \ge t)$ is approximated by (2.2) with $b = \kappa_1/\rho$, $\rho = \kappa_2/\kappa_1$ and d_3 and d_4 are given by (2.3). For T'_{01} with $5 \le k \le 40$, the values of b, ρ, d_3 and d_4 are given in Table 3. Furthermore, the two-moment approximation gives $pr(T'_{01} \ge t) \approx \bar{G}_b(t/\rho)$ for $t > 0$ with b and ρ taken from Table 3.

3.2 *Approximations to the Null Distribution of* T'_{12}. In this case, $pr(T'_{12} = 0) = 1/k$ and so we consider approximations corrected for the discrete part of T'_{12}. Under H_0, the four-moment approximation for $pr(T'_{12} \ge t)$ with $t > 0$ is given by (2.5) with b, ρ, d_3 and d_4 given in Table 4. Of course, the two-moment approximation under H_0 is given by (2.8) for $t > 0$.

Table 3. Coefficients for the Two-Moment and Four-Moment Approximations to the Null Distribution of T'_{01}.

k	ρ	b	d_3	d_4
5	2.31135	1.17029	0.03366	0.04674
6	2.28329	1.54569	0.04152	0.05762
7	2.26135	1.93529	0.04899	0.06797
8	2.24357	2.33592	0.05611	0.07784
9	2.22877	2.74539	0.06289	0.08726
10	2.21619	3.16214	0.06939	0.09627
11	2.20532	3.58499	0.07560	0.10491
12	2.19580	4.01305	0.08157	0.11322
13	2.18738	4.44559	0.08732	0.12122
14	2.17986	4.88206	0.09285	0.12893
15	2.17309	5.32199	0.09818	0.13633
16	2.16694	5.76502	0.10330	0.14343
17	2.16132	6.21083	0.10821	0.15016
18	2.15616	6.65916	0.11291	0.15644
19	2.15138	7.10984	0.11729	0.16191
20	2.14693	7.56269	0.12131	0.16635
21	2.14277	8.01756	0.12495	0.16958
22	2.13884	8.47437	0.12804	0.17112
23	2.13511	8.93305	0.13041	0.16981
24	2.13155	9.39352	0.13192	0.16507
25	2.12814	9.85572	0.13251	0.15653
26	2.12485	10.31963	0.13188	0.14279
27	2.12167	10.78520	0.12983	0.12234
28	2.11859	11.25236	0.12632	0.09456
29	2.11559	11.72110	0.12103	0.05753
30	2.11266	12.19135	0.11386	0.01116
31	2.10982	12.66299	0.10506	-0.04644
32	2.10705	13.13600	0.09394	-0.11654
33	2.10437	13.61022	0.08100	-0.19787
34	2.10178	14.08549	0.06637	-0.29185
35	2.09929	14.56163	0.05016	-0.39810
36	2.09694	15.03839	0.03309	-0.51275
37	2.09475	15.51533	0.01625	-0.63262
38	2.09273	15.99235	-0.00054	-0.75338
39	2.09096	16.46856	-0.01429	-0.86484
40	2.08930	16.94505	-0.02789	-0.97903

4. NUMERICAL COMPARISONS. For $k = 5, 10, 15$ and 20 and t successive integers, the exact value of $pr(T_{01} \geq t)$ under H_0, the two-moment, the corrected two-moment, the four-moment and the corrected four-moment approximations were computed. Table 5 gives

Table 4. Coefficients for the Corrected Two-Moment and Four-Moment Approximations to the Null Distribution of T'_{12}.

k	ρ	b	d_3	d_4
5	2.32655	0.69580	0.01390	0.01740
6	2.38563	0.73980	0.01715	0.02166
7	2.43385	0.77829	0.02001	0.02548
8	2.47431	0.81255	0.02258	0.02893
9	2.50898	0.84349	0.02492	0.03209
10	2.53917	0.87172	0.02706	0.03500
11	2.56582	0.89770	0.02905	0.03771
12	2.58961	0.92178	0.03089	0.04025
13	2.61103	0.94424	0.03262	0.04262
14	2.63048	0.96528	0.03424	0.04486
15	2.64827	0.98509	0.03576	0.04698
16	2.66463	1.00380	0.03721	0.04899
17	2.67978	1.02153	0.03858	0.05091
18	2.69386	1.03840	0.03988	0.05273
19	2.70701	1.05447	0.04112	0.05446
20	2.71935	1.06982	0.04230	0.05612
21	2.73096	1.08452	0.04342	0.05771
22	2.74193	1.09861	0.04450	0.05924
23	2.75232	1.11216	0.04553	0.06070
24	2.76220	1.12520	0.04653	0.06210
25	2.77160	1.13777	0.04748	0.06345
26	2.78058	1.14990	0.04839	0.06476
27	2.78916	1.16163	0.04928	0.06601
28	2.79739	1.17298	0.05013	0.06722
29	2.80529	1.18397	0.05095	0.06839
30	2.81288	1.19463	0.05175	0.06953
31	2.82020	1.20497	0.05253	0.07062
32	2.82725	1.21502	0.05328	0.07169
33	2.83406	1.22479	0.05401	0.07272
34	2.84064	1.23429	0.05472	0.07373
35	2.84701	1.24353	0.05541	0.07471
36	2.85319	1.25254	0.05609	0.07567
37	2.85918	1.26130	0.05675	0.07661
38	2.86501	1.26985	0.05740	0.07752
39	2.87068	1.27817	0.05804	0.07842
40	2.87618	1.28630	0.05866	0.07930

these values to four decimal places along with the percentage errors to the nearest 1/10 of a percent for $k = 5$, 10 and 20 and those t which make the exact values closest to 0.2, 0.1, 0.05, 0.01 and 0.005.

Examining Table 5, one sees that the correction for the discrete part is worthwhile. Even for k as large as 20 this is true in the right tail. For practical purposes the corrected two-moment approximation could be used except possibly for the far right tail, say at the 99th percentile and beyond. There was considerable improvement obtained by using the corrected four-moment approximation for such values for all k studied.

Similar computations were carried out for $pr(T_{12} \geq t)$ and the results are summarized in Table 6. While the trend observed in the approximation of $pr(T_{01} \geq t)$ continues in this case, it seems that for $k \geq 10$ the two-moment approximation would be adequate for practical purposes.

Studying Tables A.5 and A.6 of Barlow et al. (1972) we see $Q(\ell,k)$ behaves somewhat like $P(k-\ell+1,k)$, and so one would expect that the behavior of the approximations for $pr(T'_{01} \geq t)$ would be like those for $pr(T_{12} \geq t)$ and those for $pr(T'_{12} \geq t)$ would behave like those for $pr(T_{01} \geq t)$. For this reason we did not conduct as thorough a study of the approximations for $pr(T'_{01} \geq t)$ and $pr(T'_{12} \geq t)$. However, for $k = 10$ we did compute $pr(T'_{01} \geq t)$ for $t = 15$ and 21, as well as the two-moment and four-moment approximations. The error percentages for $t = 15$ (21) are 0.5% (8.1%) for

Table 5. Exact and Approximate Values for $pr(T_{01} \geq t)$ Under H_0.

t	two-moment	%-error	corrected two-moment	%-error	four-moment	%-error	corrected four-moment	%-error	exact
					k = 5				
2	0.2114	6.7	0.2221	2.0	0.2136	5.8	0.2219	2.1	0.2267
4	0.0786	6.0	0.0815	2.5	0.0853	2.0	0.0842	0.8	0.0836
5	0.0495	3.3	0.0503	1.7	0.0543	6.0	0.0526	2.7	0.0512
8	0.0132	10.9	0.0123	3.6	0.0128	7.3	0.0124	4.5	0.0119
10	0.0056	25.0	0.0049	9.1	0.0044	3.1	0.0045	1.0	0.0045
					k = 10				
3	0.2151	6.4	0.2219	3.4	0.2211	3.7	0.2244	2.3	0.2297
5	0.0931	5.1	0.0949	3.3	0.1014	3.4	0.1000	1.9	0.0981
7	0.0414	0.3	0.0411	0.4	0.0449	8.8	0.0438	6.0	0.0413
10	0.0126	14.8	0.0119	8.0	0.0115	4.2	0.0115	4.1	0.0110
12	0.0058	28.5	0.0052	16.1	0.0041	9.7	0.0043	5.7	0.0045
					k = 20				
4	0.2156	5.9	0.2196	4.1	0.2243	2.0	0.2251	1.7	0.2290
6	0.1029	4.5	0.1039	3.5	0.1122	4.2	0.1108	2.9	0.1077
8	0.0493	0.3	0.0490	0.3	0.0534	8.6	0.0526	6.9	0.0492
12	0.0114	18.2	0.0109	12.7	0.0096	0.6	0.0097	0.9	0.0097
14	9.9955	31.6	0.0051	22.4	0.0035	17.1	0.0036	12.8	0.0042

Table 6. Exact and Approximate Values for $pr(T_{12} \geq t)$ Under H_0.

t	two-moment	%-error	four-moment	%-error	exact
			k = 5		
4	0.2266	2.9	0.2319	0.6	0.2334
6	0.1002	1.8	0.1055	3.3	0.1021
8	0.0438	1.2	0.0456	5.5	0.0432
11	0.0125	8.7	0.0114	0.5	0.0115
13	0.0054	15.3	0.0041	11.9	0.0047
			k = 10		
10	0.2000	1.1	0.2039	0.9	0.2022
12	0.1114	0.4	0.1134	1.3	0.1119
15	0.0430	1.6	0.0426	0.5	0.0424
19	0.0110	6.1	0.0099	4.0	0.0104
21	0.0054	9.0	0.0046	7.0	0.0049
			k = 20		
21	0.2021	0.4	0.2034	0.3	0.2029
24	0.1064	0.0	0.1066	0.2	0.1063
27	0.0520	0.8	0.0515	0.2	0.0516
33	0.0104	3.4	0.0099	1.4	0.0101
35	0.0058	4.5	0.0055	1.7	0.0056

the four-moment approximation, and 1.9% (10.5%) for the two-moment approximation. These percentages are very similar to those for $pr(T_{12} \geq t)$. For $pr(T'_{12} \geq t)$, we computed the exact value, the corrected two-moment and corrected four-moment approximations for t = 7 and 12. The error percentages for t = 7 (12) are 6.0% (4.5%) for the four-moment approximation, and 0.7% (15.8%) for the two-moment approximation. Again, these percentages are much like those for $pr(T_{01} \geq t)$.

5. **SUMMARY.** When testing the homogeneity of k normal means under the assumption that they satisfy an order restriction and when testing this order restriction as the null hypothesis, the likelihood ratio statistics have chi-bar-squared null distributions provided the variances are known. Determining p-values for these tests can be tedious for moderate and large k. The level probabilities and k-1 chi-square probabilities must be computed. For the total ordering and the simple tree ordering, the constants needed to implement the two-moment and four-moment approximations discussed here are tabled. (To use the two-moment (four-moment) approximation one needs to compute one (five) chi-square probabilities.)

In the totally ordered case, correcting for the discrete part of the distribution is recommended if homogeneity is the null hypothesis, but it is not necessary if the order restriction is the null hypothesis. If homogeneity is the null hypothesis, the two-moment approximation should be used except in the far right tail (i.e., beyond the 99th percentile) and the four-moment approximation is recommended there. If the order restriction is the null hypotehsis, then for $k \geq 10$ the two-moment approxiamtion seems adequate for the significance levels considered here.

In the case of a simple tree with homogeneity as the null hypothesis, correction for the discrete part is not necessary and the two-moment approximation is adequate for practical purposes if $k \geq 10$. However, for the simple tree with the order restriction as the null hypothesis correcting for the discrete part is recommended, the two-moment approximation should be used to the left of the $99^{\underline{th}}$ percentile and the four-moment approximation used on the right of the $99^{\underline{th}}$ percentile.

REFERENCES

Barlow, R.E., Bartholomew, D.J., Bremner, J.M. & Brunk, H.D. (1972). *Statistical Inference Under Order Restrictions*, New York: Wiley.

Bartholomew, D.J. (1959). A test of homogeneity for ordered alternatives II, *Biometrika* **46**, 328-335.

Davis, A.W. (1976). Statistical distributions in univariate and multivariate Edgeworth populations, *Biometrika* **63**, 661-670.

Gideon, R.A. & Gurland, J.G. (1977). Some alternative expansions of the distribution function of a noncentral chi-square random variable, *SIAM J. Math. Anal.* **8**, 100-110.

Hirotsu, C. (1979). An F approximation and its applications, *Biometrika* **66**, 577-584.

Kotz, S., Johnson, N.L. & Boyd, D.W. (1967a). Series representations of distributions of quadratic forms in normal variables I. Central case, *Ann. Math. Statist.* **38**, 823-827.

Kotz, S., Johnson, N.L. & Boyd, D.W. (1979b). Series representations of distributions of quadratic forms in normal variables II. Non-central case, *Ann. Math. Statist.* **38**, 838-848.

Magel, R. & Wright, F.T. (1984). Tests for and against trends among Poisson intensities. In *Inequalities in Statistics and Probability: IMS Lecture Notes—Monograph Series* 5, Ed. Y.L. Tong, pp. 236-243. Hayward, California: Institute of Mathematical Statistics.

Robertson, T. (1978). Testing for and against an order restriction on multinomial parameters, *J. Amer. Statist. Assoc.* **73**, 197-202.

Robertson, T. & Wegman, E.J. (1978). Likelihood ratio tests for order restrictions in exponential families, *Ann. Statist.* **6**, 485-505.

Robertson, T. & Wright, F.T. (1983). On approximation of the level probabilities and associated distributions in order restricted inference, *Biometrika* **70**, 597-606.

Robertson, T. & Wright, F.T. (1985). One-sided comparisons for treatments with a control, *Canad. J. Statist.* **13**, 109-122.

Roy, J. & Tiku, M.L. (1962). A Laguerre series approximation to the sampling distribution of the variance, *Sankhyā* **A24**, 181-184.

Sasabuchi, S. & Kulatunga, D.D.S. (1985). Some approximations for the null distribution of the $\bar{\bar{E}}^2$ statistic used in order restricted inference, *Biometrika* **72**, 476-480.

Shirley, E. (1977). A non-parametric equivalent of Williams test for contrasting increasing dose levels of treatment, *Biometrics* **33**, 386-389.

Tan, W.Y. & Wong, S.P. (1977). On the Roy-Tiku approximation to the distribution of sample variance from nonnormal universe, *J. Amer. Statist. Assoc.* **72**, 875-881.

Tan, W.Y. & Wong, S.P. (1978). On approximating the central and noncentral multivariate gamma distributions, *Comm. Statist.* **B7**, 227-243.

Tan, W.Y. & Wong, S.P. (1980). On approximating the null and non null distributions of the F-ratio in unbalanced random effect models from nonnormal universes, *J. Amer. Statist. Assoc.* **75**, 655-662.

Tiku, M.L. (1964). Approximating the general nonnormal variance-ratio sampling distributions, *Biometrika* **51**, 83-95.

Tiku, M.L. (1965). Laguerre series forms of noncentral χ^2 and F-distributions, *Biometrika* **52**, 415-427.

Tiku, M.L. (1971). Power function of the F-test under nonnormal situations, *J. Amer. Statist. Assoc.* **66**, 913-916.

Tiku, M.L. (1975). Laguerre series forms of the distributions of classical test-statistics and their robustness in nonnormal situations, *Applied Statistics* (ed. R.P. Gupta), New York: American Elsevier Publ. Co.,

Wright, F.T. & Tran, T. (1985). Approximating thc level probabilities in order restricted inference: The simple tree bordering, *Biometrika* **72**, 429-439.

Bahadur Singh
Department of Mathematical Sciences
Memphis State University
Memphis, TN 38152

F. T. Wright
Department of Mathematics & Statistics
University of Missouri at Rolla
Rolla, MO 65401

CONDITIONAL TESTS WITH AN ORDER RESTRICTION AS A NULL HYPOTHESIS[1]

Peter C. Wollan
Michigan Technological University

Richard L. Dykstra
University of Iowa

AMS 1980 subject classifications: *Primary* 62F03; *Secondary* 62H15, 62E20.

Keywords and phrases: *isotonic inference, chi-bar-square distributions, conditional likelihood ratio tests.*

ABSTRACT

For the isotonic normal means problem, Bartholomew (1961) discussed a conditional likelihood-ratio test of H_0 : the means are homogeneous, vs. H_1 : the means satisfy the linear order. He concluded that the conditional test was substantially less powerful than the chi-bar-squared test. However, for testing H_1 vs. H_2 : all alternatives, the corresponding conditional test can be more powerful than the chi-bar-square test. Moreover, the conditional test can be modified so as to be asymptotically α-similar.

These conditional tests are of particular interest in general tests of simultaneous inequality constraints on parameters of asymptotically normal distributions, for which the coefficients corresponding to the $p(\ell,k)$'s are difficult to obtain. In this general context, the likelihood ratio statistic is asymp totically chi-bar-squared whenever the true parameter vector lies in H_1; we outline a new proof based on Silvey's theorem that a constrained estimate and its corresponding vector of Lagrange multipliers are asymptotically normal and independent.

1. This research was sponsored in part by the Office of Naval Research under ONR Contract N00014-83-K-0249.

1. INTRODUCTION. A class of problems which has received a great deal of attention involves testing hypotheses which involve order restrictions on the means of normal populations. To be specific, suppose that $\bar{x}_i$ denotes the sample mean of a random sample from a $N(\theta_i, 1)$ population, $i = 1, \cdots, k$. We assume that we have independent samples where the ith sample is of size n_i. We wish to consider the three hypotheses H_0, H_1, and H_2 defined as follows:

$$H_0 : \theta_1 = \theta_2 = \cdots = \theta_k,$$

$$H_1 : \theta_1 \le \theta_2 \le \cdots \le \theta_k,$$

$$H_2 : \text{no restrictions on the } \theta_i.$$

The problem of testing H_0 vs. H_1 was considered by Bartholomew (1961). He constructed the likelihood ratio test for this problem, which turned out to be of the form: reject H_0 for large values of the statistic

$$LR_{01} = \sum_{i=1}^{k} (x_i^* - \bar{\bar{x}}_i)^2 n_i,$$

where $x^* = (x_1^*, x_2^*, \cdots, x_k^*)$ is the maximum likelihood estimator (MLE) of $\theta = (\theta_1, \cdots, \theta_k)$ under H_1, and $\bar{\bar{x}} = (\sum_1^k n_i \bar{x}_i / \sum_1^k n_i) \cdot (1, 1, \cdots, 1)$ is the MLE under H_0. The distribution of LR_{01} under H_0 is given by

$$P(LR_{01} \ge t) = \sum_{\ell=1}^{k} P_0(\ell, k) P(\chi^2_{\ell-1} \ge t),$$

where χ_i^2 denotes a chi-square random variable with i degrees of freedom (χ_0^2 is identically zero), and the $p_0(\ell,k)$ are defined to be the probabilities under H_0 of obtaining ℓ distinct values among the x_i^*. A distribution of this form is known as a chi-bar-square, and critical values are easily obtained if the coefficients $p_0(\ell,k)$ are known. They are easy to calculate if $n_1 = n_2 = \cdots = n_k$, but very difficult for other situations.

Bartholomew also discussed a related procedure which largely avoids this difficulty. The idea, which he attributed to Tukey, was to condition on ℓ, the number of distinct values in x^*, and to compare LR_{01} to a critical value for a chi-square with degrees of freedom determined by ℓ. This is valid, because under H_0 the conditional distribution of LR_{01} given ℓ is equal to the restricted distribution, and is a chi-square with $\ell-1$ degrees of freedom (this is implicit in the proof of Theorem 3.1, Barlow et al. (1972)). In order to obtain a size α test, one must allow for the fact that $LR_{01} = 0$ with probability $p_0(1,k)$, and adjust the chi-square critical values accordingly; the form of the test becomes: reject H_0 if $LR_{01} > t_\ell$, where t_ℓ satisfies

$$P(\chi_{\ell-1}^2 \geq t_\ell) = \alpha/(1-p_0(1,k)).$$

While this conditional test was attractively simple, Bartholomew found that it was substantially less powerful than the likelihood-ratio test. (See Barlow et al. (1972), for a summary of Bartholomew's results and for further developments.)

However, Bartholomew's results concerned only tests of H_0 vs. H_1. The properties of the corresponding tests of H_1 vs. H_2 are quite different. The likelihood ratio test of H_1 vs. H_2 was constructed by Robertson and Wegman (1978); it takes the form: reject H_1 for large values of

$$LR_{12} = \sum_{i=1}^{k} (\bar{x}_i - x_i^*)^2 n_i .$$

The distribution of LR_{12} for arbitrary $\theta \in H_1$ is intractable, but Robertson and Wegman were able to show that the least favorable configuration (that is, the θ for which the probability of a type I error is maximized) is $\theta \in H_0$. Moreover, the distribution of LR_{12} under H_0 is again a chi-bar-square, involving the same $p_0(\ell,k)$ coefficients:

$$P(LR_{12} \geq t) = \sum_{\ell=1}^{k} p_0(\ell,k) P(\chi^2_{k-\ell} \geq t).$$

We will refer to this as the Chi-Bar-Square test.

As with LR_{01}, the conditional distribution under H_0 of LR_{12} given ℓ is a chi-square. Hence, we can construct a test of H_1 vs. H_2 which we will call the Conditional test: reject H_1 if $LR_{12} > t_\ell$, where t_ℓ satisfies:

$$P(\chi^2_{k-\ell} > t_\ell) = \alpha/(1-p_0(k,k)).$$

Here, the coefficient $p_0(k,k)$ is in a sense "least favorable": it is the probability, under H_0, that $LR_{12} = 0$ (or, that there are k distinct values in x^*), and it is easy to see that this probability is smaller for $\theta \in H_0$ than for any other $\theta \in H_1$.

This suggests the following modification of the Conditional test, which we will call the Adaptive test: reject H_1 if $LR_{12} > t_{\ell,\hat{p}}$, where $\hat{p}$ is a consistent estimator of $p_\theta(k,k)$, the true probability that $LR_{12} = 0$, and $t_{\ell,\hat{p}}$ satisfies

$$P(\chi^2_{k-\ell} > t_{\ell,\hat{p}}) = \alpha/(1-\hat{p}).$$

We show in Section 2 that both of these tests are symptotically of size α.

The Conditional and Adaptive tests of H_1 vs. H_2 may be preferred to the Chi-Bar-Square test for a reason other than their ease of computation: for some regions of H_2, the power of the Chi-Bar-Square test is quite low (though bounded away from zero). In these regions, the Conditional test is more powerful, and the Adaptive test is more powerful still.

In Section 3, we show that all three of these tests of H_1 vs. H_2 are valid in a quite general context, in which the underlying distribution is asymptotically normal and the hypotheses are determined by a finite set of smooth inequality constraints. This result is closely related to Theorem 2.1 of Shapiro (1985).

In Section 4, we present the results of a small simulation study, which suggest that the Conditional test can be significantly more powerful than the Chi-Bar-Square test, but that the adaptive test yields only slight additional improvement.

2. PROPERTIES OF THE CONDITIONAL AND ADAPTIVE TESTS. The fact that the Conditional and Adaptive tests are asymptotically of size α is a consequence of the following theorem.

THEOREM 1. *For any true parameter value* $\theta \in H_1$, *the* LR_{12}

statistic is asymptotically distributed as $\bar{\chi}^2(\theta)$, *that is, as a chi-bar-square whose coefficients depend on* θ.

This theorem follows from the more general Theorem 2 below, but also follows from the observation that, asymptotically, LR_{12} is also the likelihood ratio statistic for the test of the weakest partial order that θ satisfies, and the fact that the statistic for the test of this weaker hypothesis also has a chi-bar-square distribution.

Many of the properties of these tests are apparent from their rejection regions, as sketched in Figures 1 and 2. The rejection region for the Chi-Bar-Square test lies at a constant distance from H_1. For the Conditional test, the distance from the rejection region to H_1 depends on the dimension of the nearest face of H_1, the distance decreasing as the dimension increases. For the Adaptive

Figure 1. Rejection Regions of the Chi-Bar-Square and Conditional Tests. The dotted line represents the boundary of the rejection region of the Chi-Bar-Square test, and the dashed line that of the Conditional test.

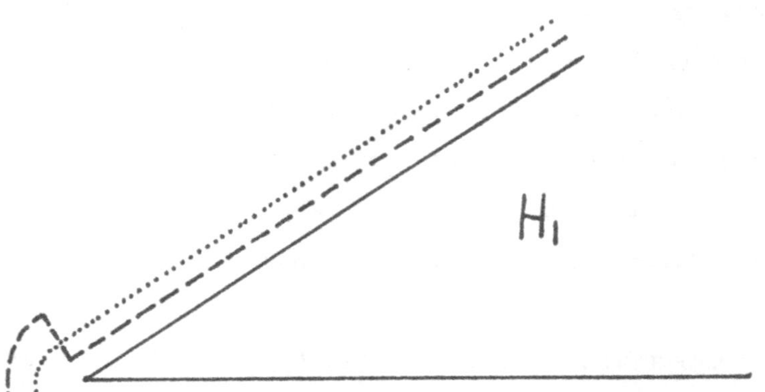

Figure 2. Rejection Regions of the Conditional and Adaptive Tests. The dotted line represents the boundary of the rejection region of the Conditional test, and the dashed line that of the Adaptive test.

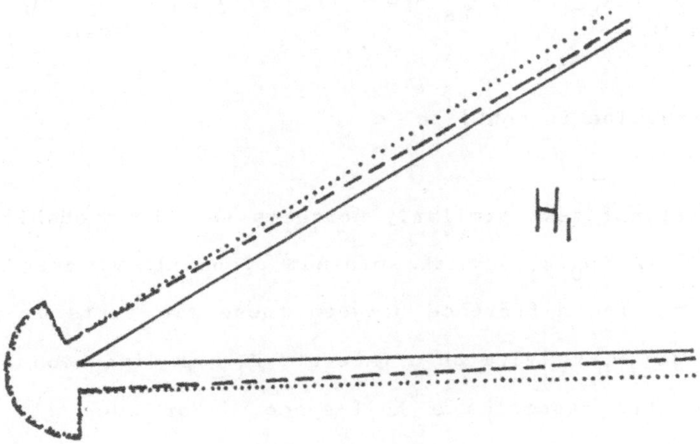

test, the actual rejection region depends on the method chosen to estimate $p_\theta(k,k)$. If the estimator is chosen to be the probability $p_{x^*}(k,k)$, corresponding to x^*, the MLE of θ restricted to be in H_1, then the boundary of the rejection region is not straight, and asymptotically approaches the boundary corresponding to a local size-α test.

In particular, the Adaptive test is asymptotically α-similar: for any θ lying on the boundary of H_1, the asymptotic probability of rejecting H_1 is α. The estimator $\hat{p}$ approaches the value $p_\theta = p_\theta(k,k)$, the asymptotic probability that $LR_{12} = 0$; but the probability of rejecting H_1 is (asymptotically)

$$\sum_{i=1}^{k} P(LR_{12} > t_{\ell, p_\theta}, \ell = i)$$

$$= \sum_{i=1}^{k-1} P(\chi^2_{k-i} > t_{i,p_\theta}) P(\ell = i) = \alpha/(1-p_\theta) \cdot \sum_{i=1}^{k-1} P(\ell = i)$$

which by construction is equal to α.

The Conditional test similarly corrects for the probability that $LR_{12} = 0$, but by $p_0(k,k)$, the minimum probability, over $\theta \in H_1$, that $LR_{12} = 0$. The difference between these two tests is greatest for θ far from H_0, lying on a face of H_1 of dimension k-1, so that H_1 locally resembles a half-space. For such a θ, the asymptotic probability of rejecting H_1 is α for the Adaptive test, and $(1/2)(\alpha/(1-p_0(k,k))) > \alpha/2$ for the Conditional test. By contrast, for the Chi-Bar-Square test, the corresponding probability is $(1/2)P(\chi^2_1 > t_k(\alpha))$, which decreases with k and can be quite small: for $\alpha = .05$, $k = 3$, it is .01620, and for $k = 4$, it is .00648.

Thus the Conditional and Adaptive tests can have substantially greater power than the Chi-Bar-Square test, for certain θ far from H_0. Near H_0, however, one must be cautious: for finite samples, the probability of rejecting H_1 using the Adaptive test will be strictly greater than using the Conditional test. The Conditional test is asymptotically of size α, so that the size of the Adaptive test may be greater than α. The extent to which this is a problem depends on the power functions of the tests, about which we know very little.

Raubertas, Lee and Nordheim (1985) have elegantly generalized

Robertson and Wegman's argument, showing that LR_{12} is stochastically minimized for $\theta \in H_0$. This does not, however, allow us to conclude that the Conditional test is of size α for finite samples, since the critical value is not constant. Nonetheless, we conjecture that for any sample size n, the Conditional test is of size α.

3. ASYMPTOTIC TESTS OF INEQUALITY CONSTRAINTS. We now consider a problem which is more general than isotonic regression. Suppose that the data have an underlying distribution depending on a k-dimensional parameter θ. Let $\{g_1,\cdots,g_m\}$ be smooth functions (m may be greater than k) and let hypotheses H_0, H_1, and H_2 be defined as follows:

$$H_0 : g_i(\theta) = 0 , \qquad i = 1,\cdots,m$$

$$H_1 : g_i(\theta) \le 0 , \qquad i = 1,\cdots,m$$

$$H_2 : \text{no restrictions.}$$

We are interested in testing H_1 vs. H_2. Let $LR_{12} = -2 \log \lambda$ be the usual likelihood ratio statistic.

Let $\mathscr{C}$ be the set of subsets of $\{1,\cdots,m\}$, and for each $c \in \mathscr{C}$ let H_c be the hypothesis determined by the set c of active constraints:

$$H_c : g_i(\theta) = 0 , \qquad i \in c.$$

Let $\hat{\theta}_c$ be the maximum likelihood estimate under H_c, and let θ^* be the MLE under H_1. For any true parameter $\theta \in H_1$, θ^* is equal to $\hat{\theta}_c$ for exactly one c (with probability one); let C be the random quantity with values in $\mathscr{C}$ such that $\hat{\theta}_C = \theta^*$.

The next theorem states that LR_{12} is asymptotically distributed as a chi-bar-square, and that one can legitimately condition on the set of active constraints. We give the proof in outline; details will appear elsewhere.

THEOREM 2. *Under suitable regularity conditions, for any* $\theta \in H_1$, $P(LR_{12} > t_C)$ *is asymptotically equal to* $\sum_{c \in C} P(\chi^{2}_{\# c} > t_c) \cdot p_c$, *where* $\chi^2_{\# c}$ *is a* χ^2 *random variable with degrees of freedom equal to the cardinality of* c, *and* p_c *is the asymptotic probability under* θ *that* $\theta^* = \theta_c$.

We have

$$P(LR_{12} > t_C) = \sum_{c \in \mathscr{C}} P(LR_{12} > t_c, \hat{\theta}_c = \theta^*).$$

It can be shown that the event $(\hat{\theta}_c = \theta^*)$ is equivalent to $(g_j(\hat{\theta}_c) < 0 \ \forall \ j \in \bar{c}, \ \hat{\lambda}_c(i) > 0 \ \forall \ i \in c)$, where $\bar{c}$ is the complement of c and $\hat{\lambda}_c(i)$ is the ith coordinate of the Lagrange multiplier vector associated with $\hat{\theta}_c$, corresponding to the constraint function $g_i(\theta)$. Moreover, if $\theta^* = \hat{\theta}_c$, then LR_{12} is asymptotically equal to a certain quadratic form in $\hat{\lambda}_c$, which we denote by LM_c. (This is one version of the Lagrange Multiplier statistic; see Engle (1983) or Buse (1982)).

Thus,

$$P(LR_{12} > t_C) \approx \sum_{c \in \mathscr{C}} P(LM_c > t_c, \hat{\lambda}_c(i) > 0 \ \forall \ i, \ g_j(\hat{\theta}_c) < 0 \ \forall \ j \in \bar{c}).$$

Silvey (1959) showed that $\sqrt{n}(\hat{\theta}_c-\theta)$ and $\sqrt{n}\,\hat{\lambda}_c$, where n is the sample size, are jointly asymptotically normal and independent. Therefore, the events $(LM_c > t_c)$ and $(g_i(\hat{\theta}_c) < 0)$ are asymptotically independent. Also, LM_c is asymptotically $\chi^2_{\#c}$, so that $(LM_c > t_c)$ is asymptotically independent of the event $(\hat{\lambda}_c(i) > 0)$, which is a function only of the direction of the vector $\hat{\lambda}_c$. Hence,

$$P(LR_{12} > t_c) \approx \sum_{c \in \mathscr{C}} P(\chi^2_{\#c} > t_c) \cdot p_c,$$

where

$$p_c = \lim P(\hat{\lambda}_c(i) > 0 \ \forall\ i,\ g_j(\hat{\theta}_c) < 0 \ \forall\ j \in \bar{c})$$

$$= \lim P(\theta^* = \hat{\theta}_c).$$

4. SIMULATION RESULTS. As we have observed, the minimum power of the Conditional test is substantially larger than the minimum power of the Chi-Bar-Square test, and the minimum power of the Adaptive test is (asymptotically) larger still. Moreover, the Chi-Bar-Square test of H_0 vs. H_1 is more powerful than the corresponding Conditional test. Thus, it is reasonable to expect that if θ were near H_0, the Chi-Bar-Square test of H_1 vs. H_2 would be the most powerful of the three, while for θ far from H_0 the Adaptive test would be most

Table 1. Estimated powers of the Chi-Bar-Squared and Conditional Tests of $H_0 : \theta = 0$ vs. $H_1 : \theta(i) \geq 0$, $i = 1,\cdots,5$, $X \sim N(\theta, I_5)$. Numbers in parentheses are two binomial standard deviations, $2\sqrt{p(1-)/1000}$.

θ	$P(LR_{01} > t)$	$P(LR_{01} > t_C)$
0	.050(.014)	.046(.013)
(1 0 0 0 0)	.127(.021)	.113(.020)
2(1 0 0 0 0)	.394(.031)	.364(.030)
4(1 0 0 0 0)	.945(.041)	.938(.015)
$(1/\sqrt{2})$(1 1 0 0 0)	.167(.024)	.134(.022)
$(2/\sqrt{2})$(1 1 0 0 0)	.430(.031)	.350(.030)
$(4/\sqrt{2})$(1 1 0 0 0)	.948(.014)	.926(.016)
$(1/\sqrt{3})$(1 1 1 0 0)	.170(.024)	.114(.020)
$(2/\sqrt{3})$(1 1 1 0 0)	.443(.031)	.331(.030)
$(4/\sqrt{3})$(1 1 1 0 0)	.962(.012)	.918(.017)
$(1/\sqrt{4})$(1 1 1 1 0)	.190(.025)	.134(.022)
$(2/\sqrt{4})$(1 1 1 1 0)	.499(.032)	.356(.030)
$(4/\sqrt{4})$(1 1 1 1 0)	.966(.011)	.919(.017)
$(1/\sqrt{5})$(1 1 1 1 1)	.202(.025)	.128(.021)
$(2/\sqrt{5})$(1 1 1 1 1)	.520(.032)	.348(.030)
$(4/\sqrt{5})$(1 1 1 1 1)	.965(.012)	.899(.019)

powerful. In Tables 1, 2, and 3, we present an exploratory simulation study, testing $H_0 : \theta = 0$ vs. $H_1 : \theta_i \geq 0$, and H_1 vs. H_2 : no restrictions, with five-dimensional $N(\theta, I)$ data. These hypotheses were chosen because they have many of the features of general inequality constraints, and yet constrained estimates are easy and the p_i coefficients are known. The results generally conform to our expectations. The Adaptive test was not carried out as described above, using an estimate of the coefficient p_ϕ corresponding to $p_\theta(k,k)$; rather, a fixed value of the coefficient was used for several values of θ. These results suggest that the Adaptive test

Table 2. Estimated powers of tests of H_1 vs. H_2, for H_1 as in Table 1 and H_2 : no restrictions.

θ	$P(LR_{12}>t)$	$P(LR_{12}>t_C)$
0	.045(.013)	.033(.011)
$0 + (1/\sqrt{5})(-1\ -1\ -1\ -1\ -1)$	.197(.025)	.113(.020)
$0 + (2/\sqrt{5})(-1\ -1\ -1\ -1\ -1)$	.538(.032)	.361(.030)
$0 + (4/\sqrt{5})(-1\ -1\ -1\ -1\ -1)$	.978(.009)	.925(.017)
$0 + (1/\sqrt{4})(0\ -1\ -1\ -1\ -1)$	.200(.025)	.132(.021)
$0 + (2/\sqrt{4})(0\ -1\ -1\ -1\ -1)$	.520(.032)	.372(.031)
$0 + (4/\sqrt{4})(0\ -1\ -1\ -1\ -1)$	.974(.010)	.923(.017)
$0 + (1/\sqrt{3})(0\ 0\ -1\ -1\ -1)$	.187(.025)	.149(.023)
$0 + (2/\sqrt{3})(0\ 0\ -1\ -1\ -1)$	.470(.032)	.377(.031)
$0 + (4/\sqrt{3})(0\ 0\ -1\ -1\ -1)$	.953(.013)	.901(.019)
$0 + 1(0\ 0\ 0\ 0\ -1)$	.145(.022)	.136(.022)
$0 + 2(0\ 0\ 0\ 0\ -1)$	.399(.031)	.379(.031)
$0 + 4(0\ 0\ 0\ 0\ -1)$	.949(.014)	.932(.016)
$(1\ 0\ 0\ 0\ 0)$	.030(.011)	.037(.012)
$(1\ 0\ 0\ 0\ 0) + (1/\sqrt{4})(0\ -1\ -1\ -1\ -1)$	.156(.023)	.125(.021)
$2(1\ 0\ 0\ 0\ 0)$	.030(.011)	.046(.013)
$2(1\ 0\ 0\ 0\ 0) + (1/\sqrt{4})(0\ -1\ -1\ -1\ -1)$	.146(.022)	.129(.021)
$4(1\ 0\ 0\ 0\ 0)$	.047(.013	.058(.015)
$4(1\ 0\ 0\ 0\ 0) + (1/\sqrt{4})(0\ -1\ -1\ -1\ -1)$	.175(.024)	.147(.022)
$(1/\sqrt{2})(1\ 1\ 0\ 0\ 0)$	.024(.010)	.035(.012)
$(1/\sqrt{2})(1\ 1\ 0\ 0\ 0) + (1/\sqrt{3})(0\ 0\ -1\ -1\ -1)$	.131(.021)	.121(.021)
$(2/\sqrt{2})(1\ 1\ 0\ 0\ 0)$	.013(.007)	.029(.011)
$(2/\sqrt{2})(1\ 1\ 0\ 0\ 0) + (1/\sqrt{3})(0\ 0\ -1\ -1\ -1)$	.135(.022)	.152(.023)
$(4/\sqrt{2})(1\ 1\ 0\ 0\ 0)$	.014(.007)	.039(.012)
$(4/\sqrt{2})(1\ 1\ 0\ 0\ 0) + (1/\sqrt{3})(0\ 0\ -1\ -1\ -1)$	.121(.021)	.157(.023)
$(1/\sqrt{2})(1\ 1\ 0\ 0\ 0) + (0\ 0\ 0\ 0\ -1)$	.085(.018)	.096(.019)
$(4/\sqrt{2})(1\ 1\ 0\ 0\ 0) + (0\ 0\ 0\ 0\ -1)$	.081(.017)	.138(.022)
$(1/\sqrt{3})(1\ 1\ 1\ 0\ 0)$	.021(.009)	.029(.011)
$(1/\sqrt{3})(1\ 1\ 1\ 0\ 0) + (1/\sqrt{2})(0\ 0\ 0\ -1\ -1)$	.113(.020)	.136(.022)
$(1/\sqrt{3})(1\ 1\ 1\ 0\ 0) + (0\ 0\ 0\ 0\ -1)$	.083(.017)	.116(.020)

θ	$P(LR_{12}>t)$	$P(LR_{12}>t_C)$
$(2/\sqrt{3})(1\ 1\ 1\ 0\ 0)$	.011(.007)	.030(.011)
$(2/\sqrt{3})(1\ 1\ 1\ 0\ 0) + (1/\sqrt{2})(0\ 0\ 0\ -1\ -1\ )$	.107(.020)	
.144(.022)$(4/\sqrt{3})(1\ 1\ 1\ 0\ 0)$	.004(.004)	.030(.011)
$(4/\sqrt{3})(1\ 1\ 1\ 0\ 0) + (1/\sqrt{2})(0\ 0\ 0\ -1\ -1)$	.073(.016)	.152(.023)
$(1/\sqrt{4})(1\ 1\ 1\ 1\ 0)$	.018(.008)	.030(.011)
$(1/\sqrt{4})(1\ 1\ 1\ 1\ 0) + (0\ 0\ 0\ 0\ -1)$	.062(.015)	.101(.019)
$(2/\sqrt{4})(1\ 1\ 1\ 1\ 0)$	.010(.006)	.025(.010)
$(2/\sqrt{4})(1\ 1\ 1\ 1\ 0) + (0\ 0\ 0\ 0\ -1)$	.060(.015)	.135(.022)
$(4/\sqrt{4})(1\ 1\ 1\ 1\ 0)$	.004(.004)	.020(.009)
$(4/\sqrt{4})(1\ 1\ 1\ 1\ 0) + (0\ 0\ 0\ 0\ -1)$	.040(.012)	.151(.023)
$(4/\sqrt{4})(1\ 1\ 1\ 1\ 0) + 2(0\ 0\ 0\ 0\ -1)$	.233(.027)	.496(.032)

remains reasonably well-behaved even if there is a large error in the estimate of p_ϕ.

5. CONCLUSION. We have discussed three tests of H_1 vs. H_2, all based on the likelihood ratio statistic LR_{12}. The Chi-Bar-Square test takes the usual form of a likelihood ratio test, so its properties are familiar. Some evidence suggests that if θ^*, the MLE under H_1, is near H_0, then the Chi-Bar-Square test is best, at least of these three. However, it performs poorly far from H_0, near the high dimension faces of H_1, and the Conditional and Adaptive test appear to have significantly greater power there.

Both the Conditional and Adaptive tests have the unsettling property that a small perturbation in the data can yield a discontinuous jump in the p-value of the test. Another aspect of this effect is that the power of the test is greater in some

Table 3. Estimated powers of the Conditional test of H_1 vs. H_2 with estimated p_ϕ. The sample p_ϕ is the observed probability that no constraints are active, and is shown only for values of θ on the boundary of H_1. Critical values were computed based on the indicated "conditional" values of p_ϕ.

θ	Sample p_ϕ	$p(LR_{12}>t_C\|p_\phi=.1)$	$P(LR_{12}>t_C\|p_\phi=0)$
$(1/\sqrt{2})(1\ 1\ 0\ 0\ 0)$	.068	.048(.014)	.035(.012)
$(1/\sqrt{2})(1\ 1\ 0\ 0\ 0)+(1/\sqrt{3})(0\ 0\ -1\ -1\ -1)$		.155(.023)	.121(.021)
$(2/\sqrt{2})(1\ 1\ 0\ 0\ 0)$	.104		
$(2/\sqrt{2})(1\ 1\ 0\ 0\ 0)+(1/\sqrt{3})(0\ 0\ -1\ -1\ -1)$		.136(.022)	.152(.023)
$(4/\sqrt{2})(1\ 1\ 0\ 0\ 0)$	.127	.051(.014)	.039).012)
$(4/\sqrt{2})(1\ 1\ 0\ 0\ 0)+(1/\sqrt{3})(0\ 0\ -1\ -1\ -1)$		.174(.024)	.157(.023)
$(1/\sqrt{3})(1\ 1\ 1\ 0\ 0)$	.089	.047(.013)	.029(.011)
$(1/\sqrt{3})(1\ 1\ 1\ 0\ 0)+(1/\sqrt{2})(0\ 0\ 0\ -1\ -1)$		.124(.021)	.136(.022)
$(1/\sqrt{4})(1\ 1\ 1\ 1\ 0)$	.117	.033(.011)	.030(.011)
$(1/\sqrt{4})(1\ 1\ 1\ 1\ 0)+(0\ 0\ 0\ 0\ -1)$		.129(.021)	.101(.019)
		$P(LR_{12}>t_C\|p_\phi=.2)$	
$(2/\sqrt{3})(1\ 1\ 1\ 0\ 0)$	.153	.046(.013)	.030(.011)
$(2/\sqrt{3})(1\ 1\ 1\ 0\ 0)+(1/\sqrt{2})(0\ 0\ 0\ -1\ -1)$		.149(.023)	.144(.022)
$(4/\sqrt{3})(1\ 1\ 1\ 0\ 0)$	.261	.042(.013)	.030(.022)
$(4/\sqrt{3})(1\ 1\ 1\ 0\ 0)+(1/\sqrt{2})(0\ 0\ 0\ -1\ -1)$		.152(.023)	.152(.023)
$(2/\sqrt{4})(1\ 1\ 1\ 1\ 0)$	.250	.028(.010)	.025(.010)
$(2/\sqrt{4})(1\ 1\ 1\ 1\ 0)+(0\ 0\ 0\ 0\ -1)$		.140(.022)	.135(.022)
$(4/\sqrt{4})(1\ 1\ 1\ 1\ 0)$	.478	.027(.010)	.020(.009)
$(4/\sqrt{4})(1\ 1\ 1\ 1\ 0)+(0\ 0\ 0\ 0\ -1)$		.183(.024)	.151(.023)
		$P(LR_{12}>t_C\|p_\phi=.4)$	
$(4/\sqrt{4})(1\ 1\ 1\ 1\ 0)$	.465	.040(.012)	.020(.009)
$(4/\sqrt{4})(1\ 1\ 1\ 1\ 0)+(0\ 0\ 0\ -1)$		.222(.026)	.151(.023)

directions away from H_1 than in others.

Clearly, the Conditional and Adaptive tests behave somewhat differently than classical statistical procedures, and further investigation is required. However, they appear to be attractive in two cases: first, in testing H_1 vs. H_2, when the true parameter value is expected to be far from H_0; and second, in tests of either H_0 vs. H_1 or H_1 vs. H_2, when values of the coefficients $\{p(\ell,k)\}$, or more generally $\{p_c\}$, are difficult to obtain. This second case will often arise from a general test of inequality constraints, under asymptotic normality. Examples and potential uses for the Conditional or Adaptive tests have appeared in the areas of multivariate analysis (Bohrer and Francis (1972)), categorical data analysis (Agresti, Chuang and Kezouh (1984), or Goodman (1985)), and econometrics (Liew (1976) or Yancey, Judge, and Bock (1981)).

REFERENCES

Agresti, A., Chuang, C. & Kezouh, A. (1984). Order-restricted score parameters in association models for contingency tables. Manuscript.

Barlow, R.E., Bartholomew, D.J., Bremner, J.M. & Brunk, H.D. (1972). *Statistical Inference Under Order Restrictions*. Wiley: New York.

Bartholomew, D.J. (1961). A test of homogeneity of means under restricted alternatives (with discussion). *J. Roy. Statist. Soc.* **B23**, 239-281.

Bohrer, R. & Francis, G.K. (1972). Sharp one-sided confidence bounds over positive regions. *Ann. Math. Statist.* **43**, 1541-1548.

Buse, A. (1982). The likelihood ratio, Wald, and Lagrange multiplier tests: an expository note. *Amer. Statist.* **36**, 153-157.

Engle, R.F. (1983). Wald, likelihood ratio, and Lagrange multiplier tests in econometrics. In *Handbook of Econometrics, Vol. II*, (Griliches and Intriligator, eds.). North Holland.

Goodman, L.A. (1985). The analysis of cross-classified data having ordered and/or unordered categories. *Ann. Statist.* **13**, 10-69.

Liew, L.K. (1976). Inequality constrained least-squares estimation. *J. Amer. Statist. Assoc.* **71**, 746-751.

Raubertas, R.F., Lee, C.C. & Nordheim, E.V. (1986). Hypothesis tests for normal means constrained by linear inequalities (to appear in *Comm. Statist*).

Robertson, Tim & Wegman E.J. (1978). Likelihood ratio tests for order restrictions in exponential families. *Ann. Statist.* **6**, 485-505.

Shapiro, A. (1985). Asymptotic distribution of test statistics in the analysis of moment structures under inequality constraints. *Biometrika* **72**, 133-144.

Silvey, S.D. (1959). The Lagrangian multiplier test. *Ann. Math. Statist.* **30**, 389-407.

Yancey, T.A., Judge, G.G. & Bock, M.E. (1981). Testing multiple equality and inequality hypotheses in economics. *Economics Letters* **7**, 249-255.

Peter C. Wollan
Department of Mathematics
and Computer Science
Michigan Technological University
Houghton, MI 49931

Richard L. Dykstra
Department of Actuarial Science
and Statistics
The University of Iowa
Iowa City, IA 52242

Lecture Notes in Statistics

Vol. 26: Robust and Nonlinear Time Series Analysis. Proceedings, 1983. Edited by J. Franke, W. Härdle and D. Martin. IX, 286 pages. 1984.

Vol. 27: A. Janssen, H. Milbrodt, H. Strasser, Infinitely Divisible Statistical Experiments. VI, 163 pages. 1985.

Vol. 28: S. Amari, Differential-Geometrical Methods in Statistics. V, 290 pages. 1985.

Vol. 29: Statistics in Ornithology. Edited by B.J.T. Morgan and P.M. North. XXV, 418 pages. 1985.

Vol. 30: J. Grandell, Stochastic Models of Air Pollutant Concentration. V, 110 pages. 1985.

Vol. 31: J. Pfanzagl, Asymptotic Expansions for General Statistical Models. VII, 505 pages. 1985.

Vol. 32: Generalized Linear Models. Proceedings, 1985. Edited by R. Gilchrist, B. Francis and J. Whittaker. VI, 178 pages. 1985.

Vol. 33: M. Csörgő, S. Csörgő, L. Horváth, An Asymptotic Theory for Empirical Reliability and Concentration Processes. V, 171 pages. 1986.

Vol. 34: D.E. Critchlow, Metric Methods for Analyzing Partially Ranked Data. X, 216 pages. 1985.

Vol. 35: Linear Statistical Inference. Proceedings, 1984. Edited by T. Caliński and W. Klonecki. VI, 318 pages. 1985.

Vol. 36: B. Matérn, Spatial Variation. Second Edition. 151 pages. 1986.

Vol. 37: Advances in Order Restricted Statistical Inference. Proceedings, 1985. Edited by R. Dykstra, T. Robertson and F.T. Wright. VIII, 295 pages. 1986.

Batch number: 10683525

Printed by Printforce, the Netherlands